D0152285

Die Grundlehren der mathematischen Wissenschaften

in Einzeldarstellungen
mit besonderer Berücksichtigung
der Anwendungsgebiete

Band 209

Gerhard Ringel

Map Color Theorem

With 176 Figures

Springer-Verlag
New York Heidelberg Berlin 1974

Gerhard Ringel
University of California, Santa Cruz, California 95064, U.S.A.

AMS Subject Classifications (1970): 05C10, 05C15, 55A15

ISBN 0-387-06548-2 Springer-Verlag New York Heidelberg Berlin
ISBN 3-540-06548-2 Springer-Verlag Berlin Heidelberg New York

 Library of Congress Catalog Card Number 73-17986. Printed in Germany. Typesetting, printing and binding: Universitätsdruckerei H. Stürtz AG, Würzburg.

Dedicated to
Professor and Mrs. J. W. T. Youngs

Foreword

In 1890 P. J. Heawood [35] published a formula which he called the Map Colour Theorem. But he forgot to prove it. Therefore the world of mathematicians called it the Heawood Conjecture. In 1968 the formula was proven and therefore again called the Map Color Theorem. (This book is written in California, thus in American English.)

Beautiful combinatorial methods were developed in order to prove the formula. The proof is divided into twelve cases. In 1966 there were three of them still unsolved. In the academic year 1967/68 J. W. T. Youngs invited me to work with him on those three cases at Santa Cruz. Surprisingly our joint effort led to the solution of all three cases. It was a year of hard work but great pleasure. Working together was extremely profitable and enjoyable.

In spite of the fact that we saw each other every day, Ted wrote a letter to me, which I present here in shortened form:

Santa Cruz, March 1, 1968

Dear Gerhard:

Last night while I was checking our results on Cases 2, 8 and 11, and thinking of the great pleasure we had in the afternoon with the extraordinarily elegant new solution for Case 11, it seemed to me appropriate to pause for a few minutes and dictate a historical memorandum.

We began working on Case 8 on 10 October 1967, and it was settled on Tuesday night, 14 November 1967. You had a crucial idea in regard to additional adjacencies on 29 October, but I could not finish the regular part of the problem with your program. On Monday afternoon, 30 October, we sat down at my desk at about 4:30 p.m. and you developed a second "program". With that miraculous good fortune which occasionally comes to mathematicians I finished the coil diagram for the case s is even in fifteen minutes. I shall long remember our pleasure at that point (it being 5 o'clock in the evening) and how we shook hands and congratulated each other.

The situation in case s is odd took longer than I had expected. You gave me a program on Monday, 13 November, and I solved the coil on

Monday night. On Tuesday morning we decided to get together at noon with sandwiches for lunch in my office, and alas, we found a mistake! As I now recall it, you had recorded a positive current at a certain point in the permutation 0 where its negative was called for. Well do I recall how much consternation this caused. I was so disturbed by your unhappiness that I remember calling you early in the evening to tell you that not only had this sort of thing happened to me many times, but I am sure it had happened to all mathematicians, but you were as I would have been, much disappointed and inconsolate. Later that evening, Tuesday, 14 November (to be quite precise, at 10:40 p.m.) you called concerned about the possibility of awakening me, to say that with the *same* program you had mentioned, you had solved the additional adjacency parts of the problem. And thus Case 8 was settled ...

On returning from my lecture trip in the Middle West, I sat down yesterday afternoon to attack the situation working with $n=35$. I made a massive table of all the various possibilities while you were lecturing to your class yesterday, and all of a sudden things began to fit together. When you dropped into my office at about 3:30 I had the current graph and was at the point of checking it. There was no mistake in the additional adjacency part of the problem. You sat at my shoulder, and it was all O.K. At this point I had the golden hunch that the new program which settled $s=2$ would work in general. We generalized the program to all s and by 4:20 p.m. we were able to exhibit the final solution with the most elegant possible coil, a coil, in fact, which had occurred on 6 October 1964 when Gustin and I tried to find an approximate solution to the Heawood Conjecture. This was wonderful news, and we now both feel that no possible further improvement will ever be made on Case 11.

Cordially yours

J. W. T. Youngs

In the spring of 1970 I met Ted again at the Mathematical Institute in Oberwolfach, Germany, for a week. We decided to write a book on the Map Color Theorem and made a first draft of a list of contents.

In the summer of 1970 I suddenly got the shocking message that Ted had died. It happened just before we were supposed to begin the book. So I was forced to write it without his help.

Three times I gave a course about graph theory and color theorems, once in Berlin and twice in Santa Cruz. During that time a collection of exercise problems was developed. They are now included in the book.

The expert will find many improvements in comparison to the present literature. It might be that some experts would like to have this book written more abstractly and more formally. For this I have to apologize.

My main concern was to make the book easily understandable for young students.

I would like to thank Edward Landesman, Heidi Mahnke, Richard Niles and David Pengelley for careful proof reading and making many good suggestions and improvements. I thank Doris Heinsohn for perfect designing of all the figures and Rosemarie Stampfel for the excellent work she did in typing the manuscript.

I would also like to thank the National Science Foundation for supporting all the research about topics connected with this book for many years.

September 1973 Gerhard Ringel

Contents

1. Problems, Illustrations, History

In this first chapter we will use geometric intuition and geometric imagination to explain all the problems and all the results which are presented in this book. In the first chapter there will be no proofs at all. It is hoped that this will serve as an introduction for those not familiar with the subject. The proofs and the mathematical foundations of graph theory and the theory of surfaces will be presented in later chapters.

1.1. The Four Color Problem

In a geographic map on the sphere (or the plane), of Europe, for example, we say two countries are *adjacent* if they have at least one common boundary. This means that Arizona and Colorado are not adjacent because they have only a point in common. We will consider only such maps on the sphere where each country has one closed curve as a boundary. If M is an arbitrary map we say the countries of M are colorable with 4 colors if it is possible to color each country with one of these colors such that any two adjacent countries have different colors.

Mathematicians do not have an answer for the following question: *Are the countries of each given map on the sphere colorable with* 4 *colors?* This unsolved problem is called the *Four Color Problem.* It is conjectured that the answer is affirmative and this is called the *Four Color Conjecture.*

We do not make any contribution to the Four Color Problem in this book. It is still an unsolved question. What we shall solve here is the Color Problem for more complicated surfaces than the sphere or the plane.

It is not known whether anyone posed the Four Color question before 1852, when Francis Guthrie put it to his younger brother Frederick [29]. Francis, a graduate student at University College, London, noticed the sufficiency of four colors for distinguishing the counties on a map of England. The first known document on the Four Color Problem is a letter dated October 23, 1852, written by Augustus DeMorgan to his friend and colleague Sir William Rowan Hamilton.

He wrote that his student Frederick Guthrie asked him the Four Color Problem and that he was not able to solve it.

The problem was entirely neglected until 1878, when Arthur Cayley told the members of the London Mathematical Society that he had been unable to solve it. In 1879 and 1880 respectively P. G. Tait [82, 83] and the English barrister A. B. Kempe [45, 46] published "proofs" of the Four Color Conjecture. (England was overpopulated with talented amateur mathematicians in those days – a statement easily confirmed by a glance at the Proceedings of the London Mathematical Society.) The English mathematician, P. J. Heawood [35], showed in 1890 that there was a gap in Kempe's proof. The unchallenged life of Kempe's argument for a decade may be cited as evidence that mathematicians of those days were no more prone to read each other's papers than they are today. Be that as it may, Heawood was not entirely destructive in his paper. He gave the first proof for the Five Color Theorem which says that the countries of each map on the sphere can be colored by five colors.

Over the years, other mathematicians have tried to solve the four color problem to no avail, although many techniques which have subsequently found applications elsewhere have been invented in an attempt to solve it.

In 1920, Philip Franklin [19] showed that all maps of 25 or fewer regions on the sphere can be colored with four colors. In the following years this number has been raised several times by various authors. The following table shows the progress:

1920	Philip Franklin [19]	25
1926	C. N. Reynolds [62]	27
1936	Philip Franklin [21]	31
1938	C. E. Winn [97]	35
1968	Oystein Ore and Joel Stemple [60]	40

This means if somebody tries to prove that the Four Color Conjecture is not true, he must exhibit a map which is not colorable with four colors and we know now that such a counterexample must have at least 41 countries.

In recent years a new type of approach to the Four Color Problem was made by applying computers intensively. For more information about this see H. Whitney and W. T. Tutte [95]. See also the two books of Oystein Ore [59] and H. Heesch [36].

1.2. Map Color Theorem

To show that the Four Color Problem is really only a part of a more complex problem we will now consider surfaces essentially different

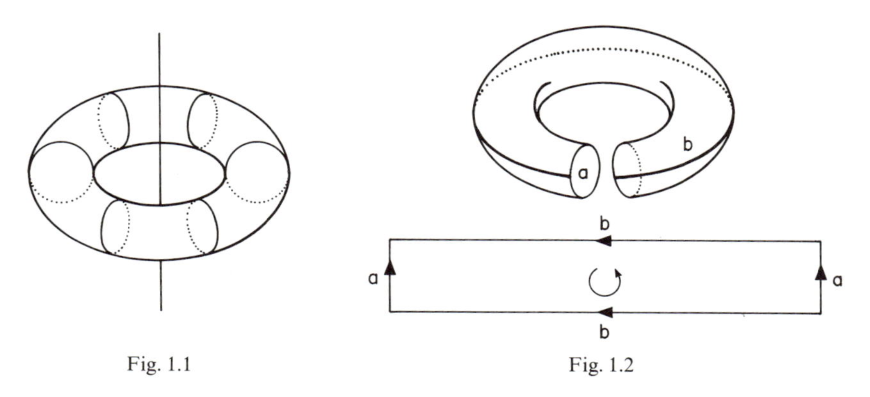

Fig. 1.1 Fig. 1.2

from the sphere. As a first example, take a circle and a non-intersecting line g. By rotating the circle around the axis g, we obtain a surface called a *torus* (Fig. 1.1). One may think of the surface of a donut or a bicycle tube. We are going to study maps on the torus. Such a study is facilitated if we first obtain a planar representation (Fig. 1.2).

Cut the torus along a generating circle a which will convert the surface into a deformed cylinder. Straighten this out and cut the cylinder along one of its altitudes b. A rectangle is then obtained. The rectangle represents the torus when the two pairs of parallel sides are identified. Fig. 1.3 shows a map with 7 countries on the torus. Each country is adjacent to every other. Fig. 1.4 gives another representation of a map on the torus with seven mutually adjacent countries. Even though each country appears to be in two parts it should be obvious how the two parts are connected on the back side of the torus.

Such a map was first described by Heawood [35]. He also proved that *each map on the torus is colorable with* 7 *colors*. The example of Figs. 1.3 and 1.4 shows that for less than 7 colors the statement is not true.

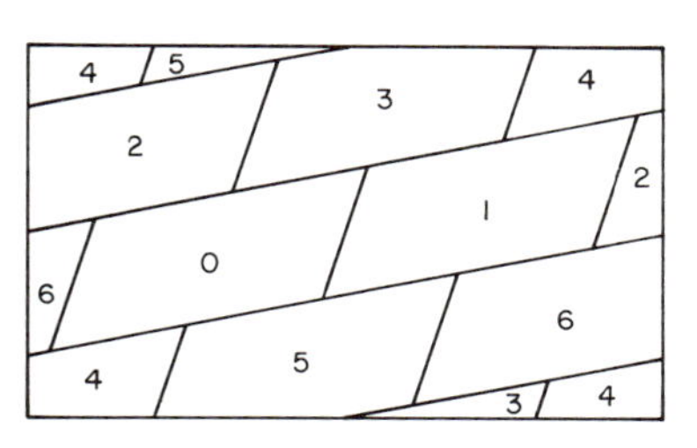

Fig. 1.3

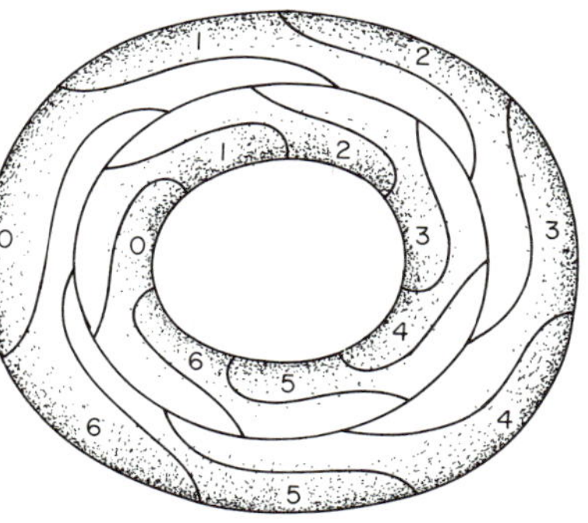

Fig. 1.4

Fig. 1.5 illustrates a sphere with 3 handles attached. More generally, let S_p be the surface of the sphere with p handles on it. (One may also think of the surface of a Swiss cheese with p holes through it.) The surface S_p is the standard model of the so-called orientable surface of genus p. Using this notation, S_0 denotes the sphere and S_1 the torus.

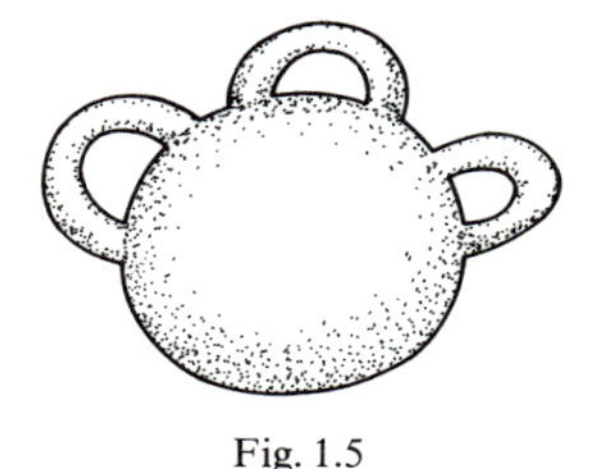

Fig. 1.5

If S is a surface and the countries of each map on S are colorable with n colors, while not every map on S is colorable with $n-1$ colors, then we say that n is the chromatic number of S and write $\chi(S)=n$. For instance, we do not know whether the chromatic number $\chi(S_0)$ of the sphere S_0 is 4 or 5, but we do know that $\chi(S_1)=7$ as has already been mentioned.

In the aforementioned paper, Heawood proved the following inequality about the chromatic number $\chi(S_p)$ on the orientable surface of genus $p \geqq 1$:

$$\chi(S_p) \leqq \left[\frac{7+\sqrt{1+48p}}{2}\right] \quad \text{for } p \geqq 1. \tag{1.1}$$

The notation $[x]$ means the largest integer not greater than x. Heawood, who wrote and proved in the occasionally casual style of the last century, was under the impression that he had shown equality in (1.1). However, it was not until 1968 that the statement

$$\chi(S_p) = \left[\frac{7+\sqrt{1+48p}}{2}\right] \quad \text{for } p \geqq 1, \tag{1.2}$$

which Heawood called the *Map Color Theorem*, was actually proved. The following table shows the chromatic number for the first few values of p.

p	1	2	3	4	5	6	7	8	9	10	11	12	13	14	15	16	17	18	19
$\chi(S_p)$	7	8	9	10	11	12	12	13	13	14	15	15	16	16	16	17	17	18	18

It is remarkable that, if the Four Color Conjecture is true, (1.1) and (1.2) hold also for $p=0$. The proof of the Eq. (1.2) has a long history. L. Heffter [37] in 1891 drew attention to the incomplete nature of Heawood's argument and was able to prove that (1.2) holds for $1 \leqq p \leqq 6$ and certain additional values of p. There the matter stood for about three quarters of a century. Before we continue with the history let us explain the following Thread Problem which is closely connected to the proof of (1.2). We shall show later that in fact the solution of the Thread Problem immediately leads to a proof of (1.2).

1.3. The Thread Problem

Suppose we are given two integers n and p with $n \geqq 3$ and $p \geqq 0$. Can we choose n points on the surface S_p and connect each of these points to each other by a simple curve (a thread) on the surface such that these curves do not intersect? This question is called the Thread Problem in the book of D. Hilbert and S. Cohn-Vossen [38].

If the answer is *yes* then we call the pair (n, p) a *possible pair*. For instance $(4, 0)$ is a possible pair because, as Fig. 1.6 shows, 4 points can be pairwise connected by mutually nonintersecting simple arcs in the plane or on the sphere. Fig. 1.7 illustrates a torus with 7 points on it mutually connected by nonintersecting arcs. Therefore $(7, 1)$ is a possible pair. The following statement is obvious: *If (n, p) is a possible pair, then the pairs $(n, p+1)$ and $(n-1, p)$ are also possible pairs.*

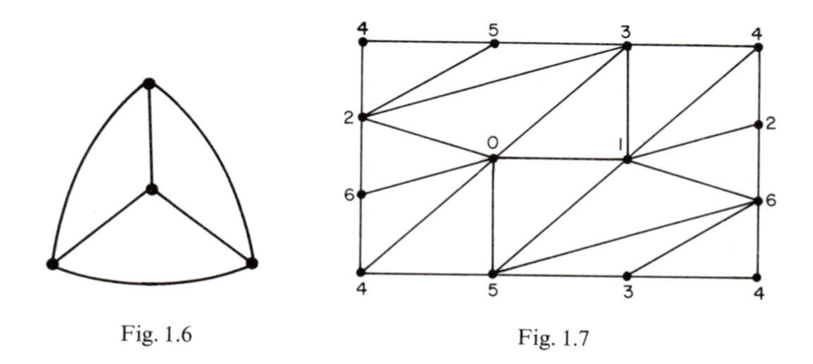

Fig. 1.6

Fig. 1.7

In order to get complete information about the set of all possible pairs we consider for any given number n the smallest possible p such that (n, p) is a possible pair. This minimal p we denote by $\gamma(n)$ because it depends on n. Since $(4, 0)$ and $(7, 1)$ are possible pairs it follows that $\gamma(4)=0$ and $\gamma(7) \leqq 1$. Moreover it can be proved that $\gamma(7)=1$. The Thread

Problem was completely solved in 1968 by proving the formula

$$\gamma(n)=\left\{\frac{(n-3)(n-4)}{12}\right\} \quad \text{for } n\geqq 3. \tag{1.3}$$

The notation $\{x\}$ means the smallest integer not less than x.

As we will see later, it is easy to prove that (1.2) is true if (1.3) is. Therefore the history of the Map Color Problem for orientable surfaces of genus $p\geqq 1$ is really the history of the progress in proving (1.3).

In 1891 L. Heffter [37] proved formula (1.3) for all $n\leqq 12$ and for the very peculiar sequence $n=19, 31, 55, 67, 139, 175, 199, \ldots$. (These are the numbers n of the form $n=12s+7$ where $q=4s+3$ is a prime number and the order of the element 2 in the multiplicative group of integers (mod q) is either $q-1$ or $(q-1)/2$.) It is not known whether this class of numbers is finite or not.

In 1952 G. Ringel [66] proved (1.3) for $n=13$ and in 1954 [67] for all $n\equiv 5 \pmod{12}$. This solution is also found in his book [71]. In 1961 he succeeded in solving the three cases $n\equiv 7, 10, 3 \pmod{12}$.

In 1963 G. Gustin [28] introduced the very powerful weapon of current graphs and announced solutions for $n\equiv 3$, 4, and 7, unaware of Ringel's solution to the first and last of these cases. Unfortunately, Gustin did not follow his research announcement with details and gave only three examples, one from each of the Cases 3, 4, and 7. It is a pity that his example in Case 4 contained a misprint.

In 1963, C. M. Terry, L. R. Welch, and J. W. T. Youngs [84] solved the case $n\equiv 0 \pmod{12}$. In 1963–65 Gustin and Youngs proved (1.3) for $n\equiv 1$ and 9 (mod 12). In 1966 Youngs [102] solved Case 6. Some of these solutions were published much later than they were discovered (see Youngs [102], [103]). In the fall of 1967 Ringel and Youngs joined forces at the University of California, Santa Cruz and solved all three remaining Cases 2, 8, and 11 in that order by the end of the year. It should be mentioned that Richard Guy helped at the end. Some of the proofs were improved in the spring of 1968.

But this is still not the end of the history. Some of the general methods break down for some small values of n. At the end of 1967 the only cases in which formula (1.3) was in doubt were $n=18, 20, 23, 30, 35, 47$, and 59. Without knowing about this situation Jean Mayer [54], Professor of French Literature at the University of Montpellier, proved (1.3) for all $n\leqq 23$ during 1967. In February 1968 Ringel and Youngs gave a joint lecture at a graph theorist meeting in Michigan. They explained what details were still missing, for instance, in case 59. One of the listeners, Richard Guy, worked the whole night and got a solution for $n=59$ by the next morning. Ringel and Youngs then solved the cases $n=35, 47$.

The last case $n=30$ was solved by Mayer at the end of February and independently by Youngs in March 1968.

The proofs of some of the cases were improved in the past years. The best success was a new elegant solution for Case 9 found by M. Jungerman [41] in 1973.

1.4. Unilateral Surfaces

In Chapter 3 we shall see that there are two classes of closed surfaces and no more. First, the orientable surface of genus p for $p=0, 1, 2, \ldots$, for which we have already considered a model (S_p) and given a formula for the chromatic number ($p \neq 0$). Second, the non-orientable surface of genus q for $q=1, 2, 3, \ldots$, for which we now do the same. Representation in three dimensional space is not easy. For the moment we will use the following model.

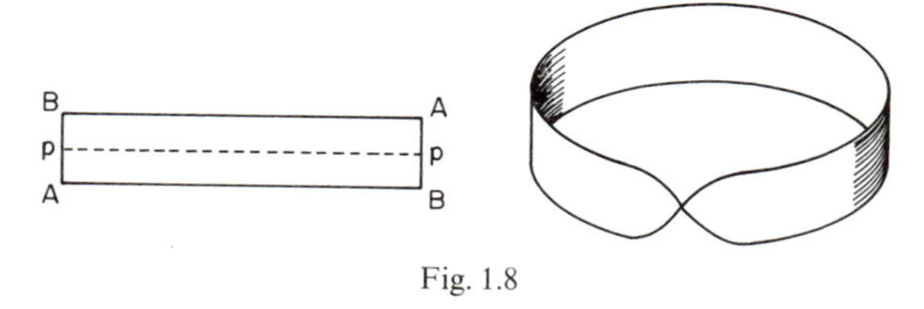

Fig. 1.8

Given a rectangle, let us label the four corners A, B, A, B in that order. Then identify the two opposite sides of the rectangle such that the first point A coincides with the second point A and likewise the two points B do the same. The resulting surface is called a *Möbius strip*, in spite of the fact that actually it was Y.B. Listing who first mentioned this surface in 1861 four years before A.F. Möbius. The boundary of the Möbius strip is *one* closed simple curve. We therefore say that the Möbius strip is a surface with boundary and not a closed surface, as is S_p. Considering the Möbius strip in 3-dimensional space we discover that it has only one side. This means that an ant traveling on the surface is able to pass, without crossing over the edge of the strip, from one side of the surface to the other. For instance, the ant can start at the midpoint P following the dotted line in Fig. 1.8. When it returns to the same point P the first time it will be on the other side of the surface. Therefore the Möbius strip is called a *unilateral* surface.

Of course on the sphere or on the sphere with p handles, the ant cannot change sides just by traveling around it. It is either inside or outside. Therefore the surface S_p is called a *bilateral* surface.

In Fig. 1.9 the two arcs labeled c should be identified so that the heads of arrows coincide. That means a Möbius strip is shown and a map on it consisting of six mutually adjacent countries. Thus $\chi(M_1) \geqq 6$ where M_1 denotes the Möbius strip. In 1910 H. Tietze [86] proved that $\chi(M_1)=6$.

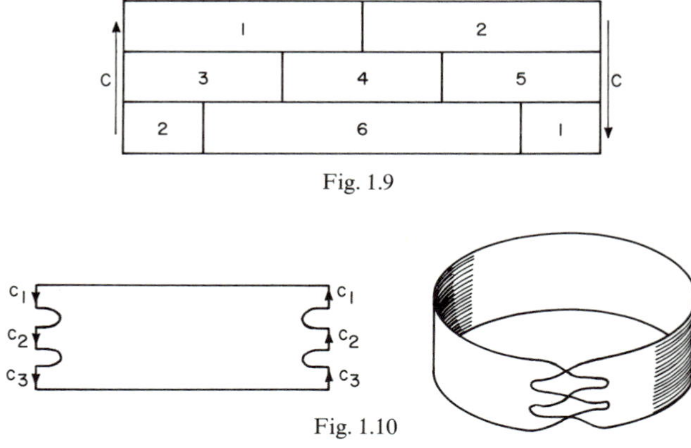

Fig. 1.9

Fig. 1.10

We now define a surface M_q which we could call a Möbius strip of order q. Take a $4q$-gon of the form illustrated in Fig. 1.10 ($q=3$). Each second arc of the $4q$-gon is labeled by $c_1, c_2, \ldots, c_q, c_q, \ldots, c_2, c_1$ in this order. Put an arrow on each of these $2q$ arcs, each pointing in the same direction on the circumference of the polygon. Then make the arc c_1 coincide with the other arc c_1 such that the heads of the arrows coincide as well. Do the same for $c_2, c_3, \ldots$. The resulting surface is denoted by M_q. It resembles a broad belt with q buckles which are all fastened after a twist of 180°. Clearly, the Möbius strip is the special case M_1. Notice that the boundary of M_q is one single closed curve.

Fig. 1.11 illustrates a map on M_2 with seven countries. Two of them, namely 3 and 5, are not adjacent. Therefore the countries of this map are colorable by six colors because the same color can be given to these two countries. In 1934 Ph. Franklin [20] proved that $\chi(M_2)=6$.

For the surface M_3 there exists a map with seven mutually adjacent countries as shown in Fig. 1.12. Therefore $\chi(M_3) \geqq 7$. Moreover it has been shown that $\chi(M_3)=7$.

The general result

$$\chi(M_q)=\left[\frac{7+\sqrt{1+24q}}{2}\right] \quad \text{for } q \neq 2, \qquad \chi(M_2)=6 \tag{1.4}$$

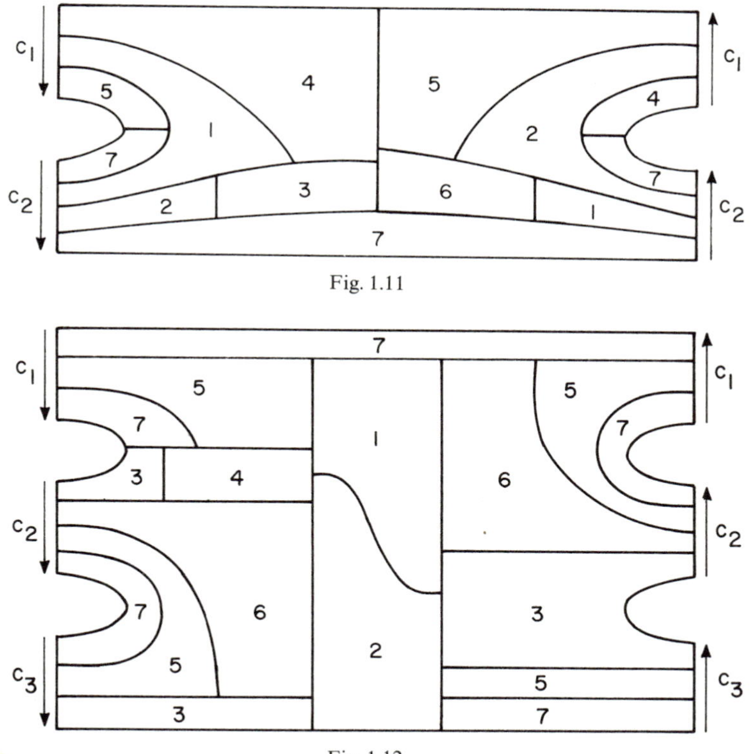

Fig. 1.11

Fig. 1.12

was proved by G. Ringel [67] in 1954. For the following special cases the formula (1.4) was proved earlier. (The first two cases have already been discussed.)

$q=1$	by H. Tietze [86] in 1910,
$q=2$	by Ph. Franklin [20] in 1934,
$q=3, 4, 6$	by I.N. Kagno [42] in 1935,
$q=5$	by H.S.M. Coxeter [8] in 1943,
$q=7$	by R.C. Bose [6] in 1939.

Ringel's proof of (1.4) is very difficult but in 1967 Youngs [100], [101] found new methods using current graphs to prove (1.4) in a much shorter and easier way. Since then some of the cases have been improved by E.M. Landesman, Youngs [49] and Ringel. It looks like the surface M_q is a very odd kind of surface. But it really can be used as a model of the non-orientable surface of genus q (see Section 3.6).

2. Graph Theory

2.1. Chromatic Number

Proper formulation of the map Color Problem and the Thread Problem and their solution requires some graph theoretical preparations.

A *graph* consists of a finite set of *vertices* and a set of *arcs* satisfying the two conditions: Each arc joins two different vertices, and no two different arcs join the same pair of vertices. For instance the 8 corners (as vertices) and the 12 edges (as arcs) of a cube form a graph.

The easiest way to describe a graph is to draw a picture. One can also give a list of all vertices and a list of all pairs of vertices joined by arcs. Fig. 2.1 illustrates some examples.

The two graphs G_6 and G_7 in Fig. 2.1 cannot be drawn in the plane without crossover (intersection) of arcs. But the important feature of a graph is its combinatorial structure. Of course in three-dimensional Euclidean space each graph can be represented such that each vertex is a point and the arcs are simple curves which do not intersect.

When, in a graph G, the two vertices A, B are joined by the arc c, we use the following terminology: A and B are each *incident* with c. We say that A and B are *adjacent* in G if there is an arc connecting A and B. The *valence* of a vertex B (denoted val B) is the number of arcs incident with B. The number of vertices (arcs) is denoted by α_0 (α_1). A vertex with valence zero is called *isolated*. The sum of the valences of all vertices in a graph counts each arc twice. Therefore we have

Theorem 2.1. *If $P_1, P_2, \ldots, P_{\alpha_0}$ are the vertices in a graph then*

$$\sum_{i=1}^{\alpha_0} \operatorname{val} P_i = 2\alpha_1$$

where α_1 is the number of arcs in G.

If H and G are two graphs and each vertex and each arc of H are contained in G then H is called a *subgraph* of G. If $H \neq G$ then H is a *proper* subgraph of G.

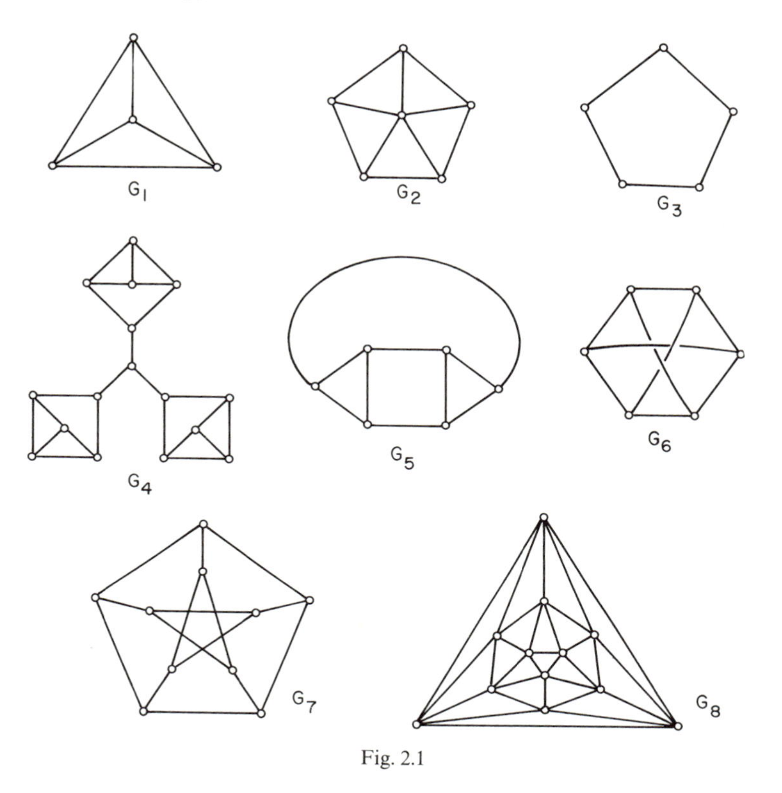

Fig. 2.1

A graph is called *n-colorable* if each vertex can be assigned one of n colors in such a way that no two vertices of the same color are adjacent. The *chromatic number* $\chi(G)$ of a graph G is the minimum number k such that G is k-colorable, i.e., $\chi(G)=k$ if G is k-colorable but not $(k-1)$-colorable.

The chromatic numbers of the first six graphs in Fig. 2.1 are 4, 4, 3, 3, 3, 2 in that order.

A graph G is called *critical* if each proper subgraph has a chromatic number smaller than that of G.

If any arc from the graph G_2 in Fig. 2.1 is removed, one gets a graph colorable with 3 colors. Therefore G_2 is critical. The graphs G_1 and G_3 in Fig. 2.1 are also critical.

Theorem 2.2. *Any graph G contains a critical subgraph H such that $\chi(H)=\chi(G)$.*

Proof. If G is critical itself, then the theorem is true for G since G is a subgraph of itself. If G is not critical then it has a *proper* subgraph H_1

such that $\chi(H_1)=\chi(G)$. If H_1 is critical then the theorem is true for $H=H_1$. If H_1 is not critical then H_1 has a *proper* subgraph H_2 with $\chi(H_2)=\chi(G)$. We continue this process. Since in each step the number of arcs or the number of vertices decreases the process stops with some subgraph H_n. Clearly then, the theorem is true with $H=H_n$.

In order to simplify notation, in the following theorems and proofs we will denote the chromatic number $\chi(G)$ of G by χ.

Theorem 2.3. *If G is critical with chromatic number χ, then the valence of each vertex of G is $\geqq \chi-1$.*

Proof. We assume that the critical graph G contains a vertex P of valence $g<\chi-1$. We label the g vertices adjacent to P by $P_1, P_2, \ldots, P_g$. We remove the vertex P and all arcs incident with P. The graph we obtain is denoted by $G-P$. The graph $G-P$ is a proper subgraph of G. Therefore $\chi(G-P)<\chi(G)$ and there is a coloration of $G-P$ using not more than $\chi-1$ colors. Consider such a coloration of $G-P$. For the g vertices $P_1, P_2, \ldots, P_g$, not more than $g \leqq \chi-2$ different colors are used. There is at least one of the $\chi-1$ colors which is not used for the g *neighbors* of P. Assigning this color to P yields a coloration of G with just $\chi-1$ colors or less, which implies that χ is not the chromatic number of G. This is a contradiction and so Theorem 2.3 is proved.

Theorem 2.4. *If G is a critical graph with α_0 vertices and α_1 arcs, and G has chromatic number χ, then the relation*

$$(\chi-1)\,\alpha_0 \leqq 2\alpha_1$$

holds.

Proof. By Theorem 2.3 each vertex of G has valence $\geqq \chi-1$. Using Theorem 2.1 one gets $(\chi-1)\,\alpha_0 \leqq 2\alpha_1$.

The *complete* graph K_n is the graph with n vertices such that each pair of vertices is joined by one arc. The graph G_1 of Fig. 2.1 is a picture of K_4. It is obvious that K_n is a critical graph with $\chi(K_n)=n$. One can check that the inequalities of Theorems 2.3 and 2.4 turn into equalities for the graph K_n.

Pseudographs. In this book we will normally only be concerned with graphs but sometimes we will need the more general concept of a pseudograph. A *pseudograph* consists of a set of *vertices* and a set of *arcs* satisfying the condition: Each arc is incident with exactly two vertices or exactly one vertex.

An arc incident with only one vertex is called a *loop*. In a pseudograph two vertices may be joined by more than one arc, as is illustrated in Fig. 2.2. Two arcs connecting the same pair of vertices are called *parallel* arcs. The *valence* of a vertex P in a pseudograph is defined as the number of arcs incident with P, but loops have to be counted twice.

For example in Fig. 2.2 val $D=8$.

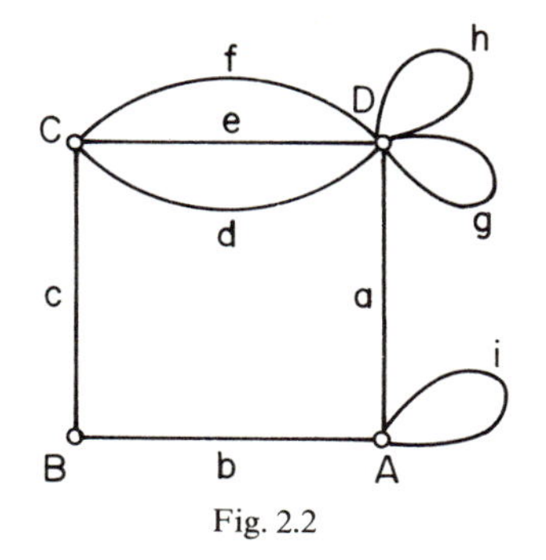

Fig. 2.2

An alternating sequence of vertices and arcs

$$P_0\, a_1\, P_1\, a_2\, P_2 \ldots a_t\, P_t \tag{2.1}$$

in a graph or in a pseudograph is called a *walk* if P_{i-1} and P_i are joined by the arc a_i $(i=1, 2, \ldots, t)$. In a *walk* a vertex or an arc may appear several times. The walk (2.1) is said to *join* vertex P_0 with vertex P_t. The *length* of a walk is the number of occurrences of arcs in it. In (2.1), the length of the walk is t.

A *trail* is a walk in which all arcs are distinct. A *way* is a walk in which all vertices (and hence all arcs) are distinct. If in (2.1) $P_0=P_t$ then

$$P_0\, a_1\, P_1\, a_2\, P_2 \ldots a_t \tag{2.2}$$

is a *closed walk*. In this case row (2.2) is read cyclically, which means that P_0 follows a_t. For example in the pseudograph of Fig. 2.2,

$$C\, f\, D\, f\, C\, d\, D\, h\, D\, a\, A$$

is a walk but not a trail. The sequence,

$$C\, f\, D\, e\, C\, d\, D\, h\, D\, a\, A$$

is a walk and a trail but not a way and

$$C\, f\, D\, a\, A$$

is a way.

In a pseudograph one may possibly find closed ways of length two or one. Both are impossible in graphs. A walk like (2.1) in a graph (not in a pseudograph) can be described in shorter notation as

$$P_0\, P_1\, P_2 \ldots P_t \quad \text{or} \quad a_1\, a_2 \ldots a_t.$$

In either case the complete alternating sequence (2.1) is uniquely determined. A graph or a pseudograph is *connected* if for each pair of vertices there exists a way leading from one to the other. A graph without any closed ways is called a *forest*. A connected forest is a *tree*.

Digraphs. If c is an arc joining the vertices A and B in a graph, then the arc c can have one of two possible orientations. An *orientation* of c is a pair (A, B) where A is the *initial* and B the *terminal* vertex of the arc. The opposite orientation is the pair (B, A). An orientation, then, is simply an assignment of a direction to an arc. An oriented arc is usually represented by placing an arrow on the arc pointing towards the terminal vertex. A loop in a pseudograph can also be oriented in two ways but it is not possible to describe this combinatorially.

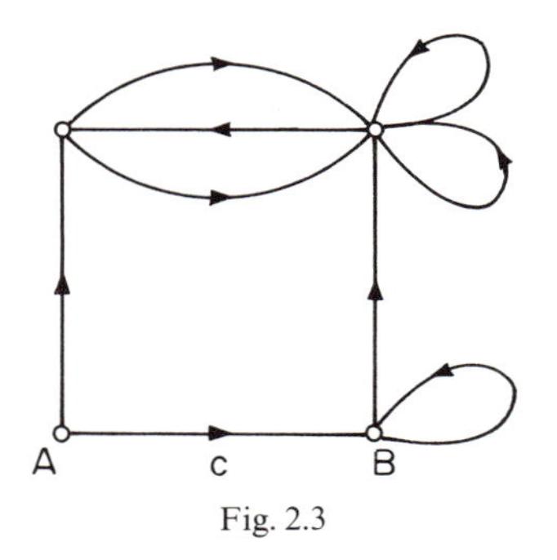

Fig. 2.3

A *directed* pseudograph (shortened to *digraph*) is simply a pseudograph where each arc is oriented. Fig. 2.3 illustrates this.

Exercises

2.1.1. Does there exist a graph with one vertex of valence five and all the others of valence four?

2.1.2. Determine the chromatic number of the two graphs G_7 and G_8 in Fig. 2.1. Find critical subgraphs with the same chromatic number.

2.1.3. What is $\chi(C_n)$ if C_n is the graph consisting of the corners (as vertices) and sides (as arcs) of an n-sided polygon?

2.1.4. Show that the number of vertices of odd valence in a graph is always even.

2.1.5. Prove the theorem: In each graph with at least one arc you can find two vertices of the same valence.

2.1.6. The complete graph K_n is critical and the equation $(\chi-1)\alpha_0=2\alpha_1$ holds for $\chi=\chi(K_n)$. (Compare with Theorem 2.4.) Are there any other critical graphs satisfying that equation?

2.1.7. Find a 2-colorable graph with $2n$ vertices and a maximal numbers of arcs. How many arcs are in that graph?

2.1.8. Prove the theorem: If each closed way in a graph is of even length, then $\chi(G)\leqq 2$.

2.1.9. The distance $d(A, B)$ of two vertices in a graph is the length of the shortest way from A to B. A tree has the property that for each pair A, B of vertices there exists only one way from A to B with length $d(A, B)$. Are there any other connected graphs with that property? Draw some examples.

2.1.10. If G is a graph, the *complement* $\bar{G}$ is defined as follows: The vertices of $\bar{G}$ are the vertices of G but any two vertices which are not adjacent in G are adjacent in $\bar{G}$ and vice versa. Show the statement: If G is not connected then $\bar{G}$ is connected.

2.1.11. Let G be a connected graph with α_0 vertices and α_1 arcs. Show that $\alpha_0\leqq\alpha_1+1$ holds and that the equation $\alpha_0=\alpha_1+1$ is true if and only if G is a tree. (Hint: Use induction and consider a circle or the longest possible way in the graph.)

2.1.12. Prove the following statement: If G is a graph and $\chi(G-A-B)=\chi(G)-2$ holds for each pair A, B of vertices of G, then G is a complete graph.

2.1.13. Take equal sized discs (use pennies, etc.) and place them flat on a table such that no two discs overlap. Define $\chi(A)$ to be the minimum number of colors needed to color each disc in an arrangement A such that no two attached discs have the same color. Prove that $\max\chi(A)=4$ where the maximum is taken over all possible configurations A. (You have to display an arrangement A with $\chi(A)=4$ and prove that $\chi(A)\leqq 4$ holds for any A.)

2.1.14. A forest has α_0 vertices and α_1 arcs. How many trees "grow" in this forest? We must explain what is meant by "grow". A subgraph H in a graph G is called a maximal connected subgraph of G if there is no connected subgraph $K\neq H$ of G such that H is a subgraph of K. A maximal connected subgraph of G is sometimes also called a *component* of G. If G is not connected, then it consists of several components. The question is: How many components does the given forest have?

2.1.15. If a tree has n arcs and one vertex is of valence n then the other vertices must have valence 1. Such a tree is called an umbrella U_n.

A tree with n arcs and the property that all vertices are of valence $\leqq 2$ is called a way W_n of length n.

A connected graph with n arcs and the property that each vertex has valence 2 is called an n-cycle C_n.

Let $n > 3$ and G be one of the graphs K_3, U_3, W_3, C_4, U_4, U_n, W_n, C_n, C_{n+1}.

The question is: How many subgraphs isomorphic to G are in the complete graph K_{n+1}?

2.2. Rotations of Graphs

The understanding of graphs with rotations and the ability to be at ease in working with them are absolutely essential in this book.

Given a graph G, a *rotation* of a vertex A of G is an oriented cyclic order (or a cyclic permutation) of all arcs incident with A. We will denote the vertices of a graph by the numbers $0, 1, 2, \ldots$. If the vertex 0 is of valence 3 and the three vertices adjacent to 0 are 1, 2, 3, then there are two different possibilities for the rotation of vertex 0. We can describe the rotation by presenting the cyclic order of the adjacent vertices of 0 instead of the arcs incident with 0. Then the two possibilities are

$$(1\,2\,3) = (2\,3\,1) = (3\,1\,2)$$

or

$$(3\,2\,1) = (2\,1\,3) = (1\,3\,2).$$

A vertex of valence 4 can have one of six different rotations. More generally the number of possible rotations of a vertex of valence n is $(n-1)! = 1 \cdot 2 \cdot 3 \ldots (n-1)$.

A *rotation* σ of a graph G is a rotation for each vertex of G. The notation (G, σ) means a graph G with a certain rotation σ.

Often it is customary to represent a graph G with rotation in the plane in such a way that a clockwise (or counterclockwise) reading of the arcs incident with a vertex gives the rotation at that vertex. More-

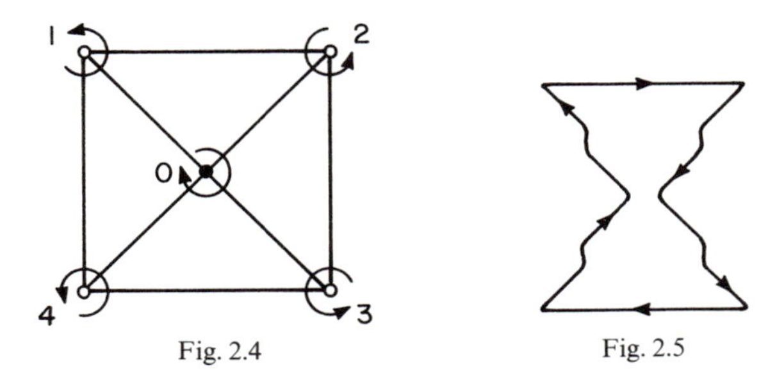

Fig. 2.4

Fig. 2.5

over, if the reading is to be clockwise (counterclockwise) the vertex is represented by a small filled-in (empty) circle such as ● (o).

In the following way one can describe and present a rotation of a graph. Denote the vertices by the numbers 0, 1, 2, ... Then write down the cyclic permutation of the neighbors of each vertex i. This permutation is given by the rotation of the vertex i, which is really a cyclic permutation of the arcs incident with i. For instance the scheme

(2.3)

0.	1	2	3	4
1.	4	0	2	
2.	0	3	1	
3.	2	0	4	
4.	3	0	1	

represents the graph with rotation of Fig. 2.4.

The diagram

(2.4)

0.	1	3	2	4
1.	3	0	2	4
2.	1	4	0	3
3.	2	4	1	0
4.	1	0	3	2

represents an interesting rotation of the complete graph K_5. It is shown in Fig. 2.6.

We consider now certain closed walks in a graph which are determined by a given rotation of the graph. In order to explain this easily, consider the graph as a road map, the arcs as two-way highways, and each vertex as a one-way traffic circle whose rotation determines the direction of the flow of traffic, i.e., clockwise or counterclockwise.

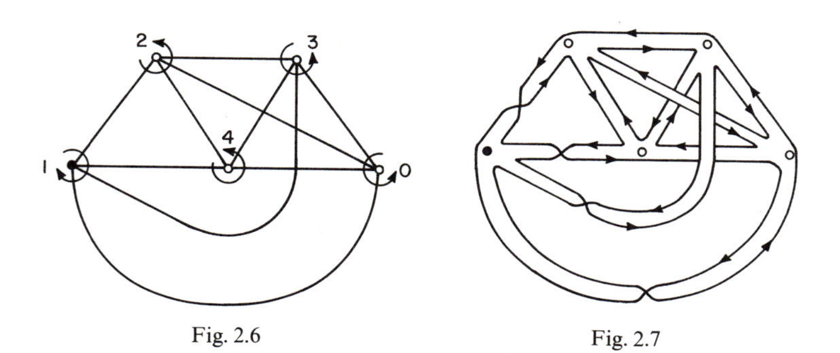

Fig. 2.6

Fig. 2.7

Assume a traveller takes a trip starting at vertex A_0, continuing on highway c_0, and each time he reaches the next vertex, regarded as a traffic circle, he takes the first exit. (This, of course, puts him on another highway if the valence of the vertex is $\geqq 2$. If the valence is one he gets back to the same highway but in the opposite direction.) We call this trip the circuit generated by A_0 and c_0 which is induced by σ. The trip is really a walk, because it may happen that vertices or arcs are repeated, although each time an arc is repeated, the repetition occurs in the opposite direction.

For instance, Fig. 2.5 illustrates the circuit 1, 2, 0, 3, 4, 0 induced by the graph with rotation of Fig. 2.4. Two other circuits are induced too: 4, 3, 2, 1 and 1, 0, 2, 3, 0, 4. The rotation of K_5 given by (2.4) or Fig. 2.6 induces one single circuit

$$1\,2\,4\,1\,3\,0\,2\,3\,4\,2\,0\,4\,3\,1\,0\,3\,2\,1\,4\,0$$

as shown in Fig. 2.7. (We have omitted the commas.) Notice that each arc appears exactly twice, the second time always in the reverse direction.

The reader may think that we have used too many undefined words like traveller, trip, traffic, etc. It is not difficult to describe the above concepts in a precise mathematical way. We do so as follows:

Let A_0 be a vertex incident with an arc c_0 in a graph G with rotation (G, σ). We are going to construct a closed walk

$$A_0, c_0, A_1, c_1, A_2, c_2, \ldots \tag{2.5}$$

in G, where A_1 is the second endpoint of c_0 and c_1 is the successor of c_0 in the rotation of vertex A_1 given by σ. A_2 is then the vertex incident with c_1 but not equal to A_1. We then take c_2 as the successor of c_1 in the rotation of A_2 and so on.

We terminate the process just before the pair A_0, c_0 is repeated. This occurs because G is finite and the process is also uniquely determined in the reverse direction, i.e., if the part $\ldots A_{t-1}, c_t, A_t \ldots$ is known, c_{t-1} is given by the rotation around A_{t-1}. We call this closed walk (2.5) the *circuit* which is generated by A_0, c_0 and *induced* by the rotation σ. We denote the number of circuits induced by a rotation σ of a graph G by $\alpha_2(G, \sigma)$. We consider the two numbers

$$\mu(G) = \max \alpha_2(G, \sigma)$$

and

$$\nu(G) = \min \alpha_2(G, \sigma)$$

where the maximum (or the minimum) is taken for all possible rotations σ of G. We will call a rotation which induces exactly $\mu(G)$ or exactly $\nu(G)$ circuits a *maximal* and a *minimal rotation* respectively.

A rotation inducing exactly one circuit is called a *circular* rotation. A circular rotation is certainly minimal. But there are graphs where a minimal rotation is not circular. See for instance G_4 in Fig. 2.1. Each rotation of a tree is circular and therefore maximal and minimal.

Suppose (G, σ) is a graph with a rotation σ. If c is an arc of G, consider the subgraph $G-c$. We define $(G-c, \sigma)$ as follows: If A is a vertex not incident with c, then the rotation of A in $(G-c, \sigma)$ is the same as the rotation of A in (G, σ). If A is incident with c, then we obtain the rotation of A in $(G-c, \sigma)$ in the obvious way, i.e., by omitting the arc c from the rotation of A in (G, σ). See Fig. 2.8.

An arc incident with a vertex of valence 1 is called a *dead-end-arc*.

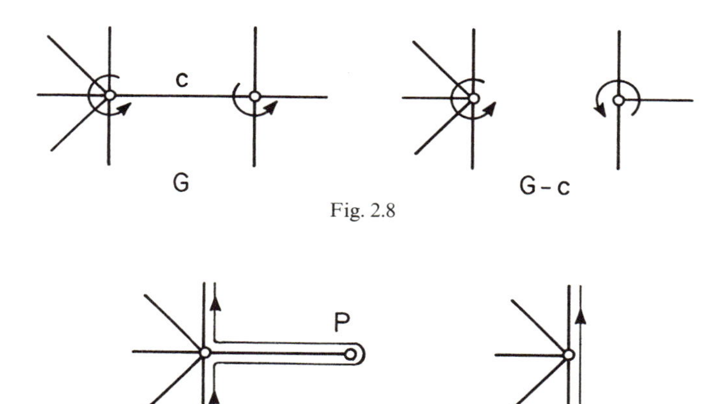

Fig. 2.8

Fig. 2.9

Theorem 2.5. *Let G be a graph with rotation (G, σ) and c an arc of G which is not a dead-end-arc. Then $\alpha_2(G, \sigma)$ differs from $\alpha_2(G-c, \sigma)$ by exactly* 1.

Proof. (See Figs. 2.10 and 2.11.) Suppose A and B are the two vertices incident with the arc c. Then removing c from G can affect at most two circuits; the circuit α, generated by A, c and the circuit β, generated by B, c. It may happen that $\alpha=\beta$ or it may happen that $\alpha \neq \beta$.

If $\alpha \neq \beta$ (Fig. 2.10) then the two circuits α and β fuse into one when c is removed. If $\alpha=\beta$ (Fig. 2.11), then the one circuit divides into two when c is removed. Hence Theorem 2.5 follows.

Recall the notation $a \equiv b \pmod{m}$, which means that $b-a$ is a multiple of m, e.g., $a \equiv 0 \pmod{2}$ means that a is an even number.

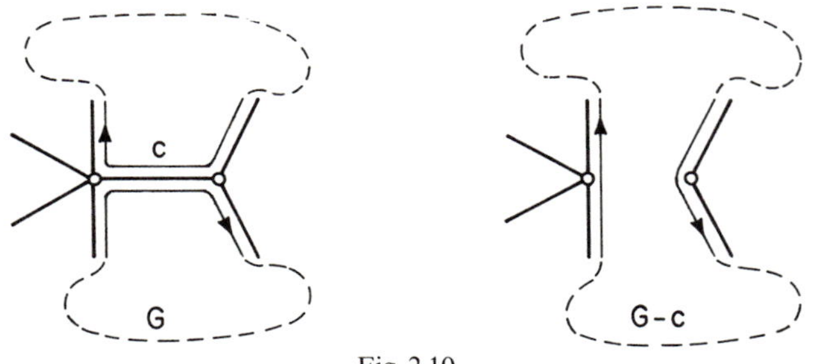

Fig. 2.10

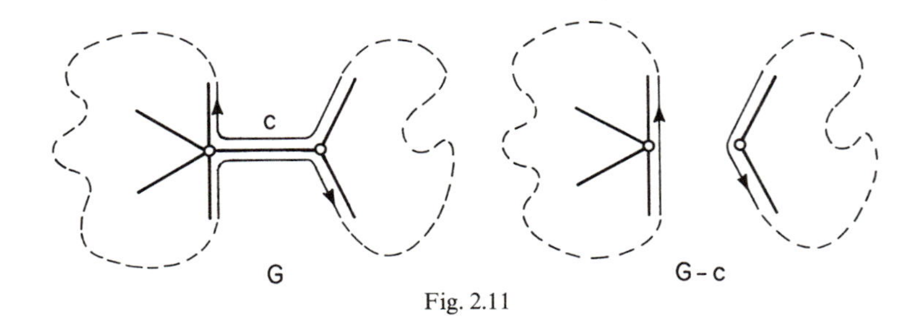

Fig. 2.11

Theorem 2.6. *If G is a graph with α_0 non-isolated vertices and α_1 arcs then*

$$\alpha_0-\alpha_1+\alpha_2(G,\sigma)\equiv 0 \pmod 2$$

holds for any rotation σ of G.

Proof. Without loss of generality suppose that G is connected. We use induction on the number of arcs in G. If G has only one or two arcs then each rotation of G induces just one circuit and Theorem 2.6 is true for G.

Now let G be a given graph and suppose for each graph with less arcs, Theorem 2.6 holds. We consider two cases:

1) If G contains a vertex P of valence one, consider the graph $G-P$ obtained by removing P and the dead-end-arc incident with P. The rotation of G can be carried over in a natural way to $G-P$. (See Fig. 2.9.) For $G-P$, Theorem 2.6 holds by the induction assumption and gives

$$(\alpha_0-1)-(\alpha_1-1)+\alpha_2(G-P,\sigma)\equiv 0 \pmod 2).$$

Obviously $\alpha_2(G,\sigma)=\alpha_2(G-P,\sigma)$ and therefore Theorem 2.6 holds for G.

2) Assume G contains an arc c which is not a dead-end-arc. By removing c we get the graph $G-c$. This graph $G-c$ has less arcs and therefore again by the induction assumption, Theorem 2.6 is true for

$G-c$. We have

$$\alpha_0-(\alpha_1-1)+\alpha_2(G-c,\sigma)\equiv 0 \pmod 2.$$

By Theorem 2.5, the circuit number $\alpha_2(G,\sigma)$ differs from $\alpha_2(G-c,\sigma)$ by one. This shows that Theorem 2.6 also holds for G.

If each circuit induced by a rotation σ for a graph has length three then σ is called a *triangular rotation*. A circuit of length three is sometimes called a *triangle*. Suppose G has no vertices of valence $\leqq 1$. Then there are no circuits shorter than triangles. Therefore a *triangular rotation* of G is a *maximal rotation*.

Later we will see that the problem of finding a maximal rotation is equivalent to the very important question of how to determine the genus of G (see Chapter 5). The theory of rotations gives us the opportunity to deal with such questions purely combinatorially, i.e., without considering any surfaces.

It is even possible to define the *combinatorial genus* of a graph G as the number p in the equation

$$\alpha_0-\alpha_1+\alpha_2(G,\sigma)=2-2p$$

for a maximal rotation σ of G. It will turn out that the combinatorial genus is really the same as the genus of G (see Chapter 5).

We are interested in finding triangular or maximal rotations of the complete graph K_n. If one takes the clockwise rotation for each vertex in the graph G_1 of Fig. 2.1, one gets a triangular rotation for K_4. The next theorem shows that one cannot expect the existence of a triangular rotation of K_n for each n.

Theorem 2.7. *If there exists a triangular rotation of the complete graph K_n then $n\equiv 0, 3, 4,$ or $7 \pmod{12}$.*

Proof. Assume that σ is a triangular rotation of K_n. Each of the triangles produced by the rotation σ uses three arcs and each arc belongs to two of these triangles. Therefore

$$3\alpha_2(K_n,\sigma)=2\alpha_1.$$

After multiplying the formula of Theorem 2.6 by six and substituting $\alpha_0=n$ and $\alpha_1=\dfrac{n(n-1)}{2}$ we get

$$6\alpha_0-6\alpha_1+6\alpha_2(K_n,\sigma)\equiv 0 \pmod{12},$$

$$6\alpha_0-2\alpha_1\equiv 0 \pmod{12},$$

$$6n-n(n-1)\equiv 0 \pmod{12},$$

$$(7-n)\,n\equiv 0 \pmod{12}. \tag{2.6}$$

There are four solutions of (2.6), namely, $n \equiv 0$, 3, 4, or 7 (mod 12). This completes the proof.

More important than Theorem 2.7 is the converse statement that for all $n \equiv 0$, 3, 4, or 7 (mod 12) a triangular rotation for K_n really exists. A great part of this book is spent dealing with it. The proof for the easiest of these four cases will be presented later in this chapter. As an introduction to the general solution we will now give some examples of graphs with rotations. Thereby we can study schemes in more detail.

We denote the seven vertices of a complete graph K_7 by 0, 1, 2, 3, 4, 5, 6. The following scheme exhibits a rotation of K_7.

(2.7)

0.	1	3	2	6	4	5
1.	2	4	3	0	5	6
2.	3	5	4	1	6	0
3.	4	6	5	2	0	1
4.	5	0	6	3	1	2
5.	6	1	0	4	2	3
6.	0	2	1	5	3	4

Of course the lines are read cyclically. This means, for instance, that in line 6 the number 0 follows 4.

We should not disguise the fact that we have really gotten this scheme (2.7) from the example in Fig. 1.7 illustrating an embedding of K_7 into the torus. But in this chapter we intend not to consider surfaces. Here the subject should remain one-dimensional.

One can easily check that the scheme (2.7) satisfies the following rule Δ^*, the rule of triangles:

Rule Δ^*. *If in line i one has i. ...j k... then in line k one must have k. ...i j....*

For instance one sees 1. ...2 4... and 4. ...1 2... in the scheme (2.7). Rule Δ^* guarantees that the given rotation (2.7) of K_7 is a triangular one. We can show this using Fig. 2.12. Given the part *i. ...j k...* of the scheme, this means in the rotation of the vertex *i* the arc joining *i* and *k*

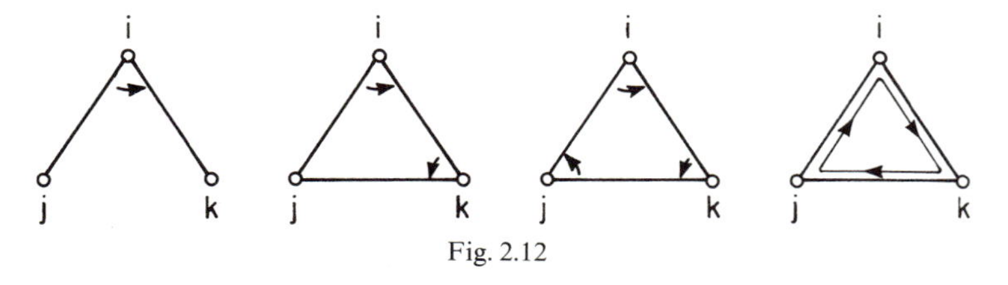

Fig. 2.12

follows the arc joining i and j (first part of Fig. 2.12). Using Rule Δ^* one gets $k. \dots i j \dots$ and therefore information on the rotation at k (second part of Fig. 2.12). When Rule Δ^* is used a second time one gets $j. \dots k i \dots$ and this closes the circuit (third and last part of the picture). Rule Δ^* is really just another expression for the fact that the rotation is triangular.

We present a triangular rotation for another graph. The notation $K_{10}-K_3$ means the graph which one gets by removing 3 arcs forming a triangle from the graph K_{10}. We denote the vertices of $K_{10}-K_3$ by 0, 1, 2, 3, 4, 5, 6, x, y, z. The rotation exhibited by

(2.8)

0.	1	x	6	2	y	5	4	z	3
1.	2	x	0	3	y	6	5	z	4
2.	3	x	1	4	y	0	6	z	5
3.	4	x	2	5	y	1	0	z	6
4.	5	x	3	6	y	2	1	z	0
5.	6	x	4	0	y	3	2	z	1
6.	0	x	5	1	y	4	3	z	2
x.	0	1	2	3	4	5	6		
y.	0	2	4	6	1	3	5		
z.	0	4	1	5	2	6	3		

satisfies rule Δ^* and consequently is triangular. One can see that the three vertices x, y, z are not adjacent to each other because in the line x there is no y or z and in line y no z.

Theorem 2.7 says that there is no triangular rotation for K_{10} and that is why we considered $K_{10}-K_3$. The following scheme is a rotation for K_8. It can not be triangular as Theorem 2.7 shows.

(2.9)

0.	2	7	3	1	4	5	6
2.	4	1	5	3	6	7	0
4.	6	3	7	5	0	1	2
6.	0	5	1	7	2	3	4
1.	7	6	5	2	4	0	3
3.	1	0	7	4	6	2	5
5.	3	2	1	6	0	4	7
7.	5	4	3	0	2	6	1

We can check that the rotation (2.9) of K_8 induces the two oriented quadrilaterals

$$0246 \quad \text{and} \quad 7531$$

and all the other circuits are oriented triangles. Since in K_8 there are 28 arcs, each covered twice by the complete set of circuits, there are 56 edges to be traveled. Each circuit has at least length 3. Therefore the number of circuits is at most one third of 56. That means $\alpha_2(K_8, \sigma) \leqq 18$ holds for each rotation σ of K_8. Therefore the rotation given by (2.9) which induces 16 triangles and 2 quadrilaterals is a maximal rotation of K_8.

Exercises

2.2.1. Does a rotation of the graph of Fig. 2.4 exist which induces just one circuit?

2.2.2. Show that each rotation of the graph G_4 in Fig. 2.1 induces four or more circuits.

2.2.3. Find a maximal rotation for the graphs G_4 and G_6 in Fig. 2.1.

2.2.4. Does a triangular rotation for the graph G_8 in Fig. 2.1 exist?

2.2.5. A graph is called *trivalent* if each vertex is of valence 3. Prove the following statement: If the number of arcs in a trivalent graph G is even (odd) then each rotation of G induces an even (odd) number of circuits.

2.2.6. Let G be a graph with the vertices $P_1, P_2 \ldots P_t$. Suppose the valence of each P_i is $v_i \geqq 2$ $(i = 1, 2, \ldots, t)$. How many different rotations of G are there?

2.2.7. Let n be a positive integer $\equiv 0$ or 3 (mod 4). Show that for the complete graph K_n there is no circular rotation.

2.2.8. Prove the following statement without using Theorems or Exercises: If G is a trivalent graph then $\alpha_2(G, \sigma) \equiv \alpha_2(G, \sigma')$ (mod 2) for any two rotations σ, σ' of G.

2.2.9. Find a triangular rotation of the graph $K_5 - K_2$.

2.2.10. What is the number of induced circuits in the rotation of K_6 given by

$$\begin{array}{llllll} 0. & 4 & 3 & 5 & 2 & 1 \\ 1. & 2 & 3 & 4 & 5 & 0 \\ 2. & 3 & 4 & 5 & 0 & 1 \\ 3. & 4 & 5 & 0 & 1 & 2 \\ 4. & 5 & 0 & 1 & 2 & 3 \\ 5. & 0 & 1 & 2 & 3 & 4 \end{array} \qquad ?$$

2.3. Orientable Cases 7 and 10

We intend to show that there exists a triangular rotation of K_n if $n \equiv 7 \pmod{12}$ and also of $K_n - K_3$ if $n \equiv 10 \pmod{12}$.

It will be convenient to label the vertices of a graph using elements of a group and possibly additional other symbols which are not group elements.

First we study the rotation for K_7 given by (2.7) and for $K_{10} - K_3$ given by (2.8). We interpret the labels 0, 1, 2, 3, 4, 5, 6 as the elements of the cyclic group $\mathbb{Z}_7$, the additive group of integers modulo 7. The symbols x, y, z are not considered to be group elements.

We observe that in the schemes (2.7) and (2.8) *all the information* for constructing them is carried in row 0. The other rows are "manufactured" by the

Additive Rule. *To obtain row i add i to each element in row* 0 *with no change in order. Addition is carried out in the group Γ which is used to represent the vertices. If there is a symbol $z \notin \Gamma$ in row* 0, *define $z + i = z$.*

The additive rule does not give a recipe for finding row z. *However row z can be manufactured by using rule Δ^*.*

A scheme obtained from row 0 using these two principles is said to be of index 1.

Theorem 2.8. *For each positive integer $n \equiv 7 \pmod{12}$ there exists a triangular rotation of the complete graph K_n.*

For every n of the form $n = 12s + 7$ we want to find a scheme satisfying rule Δ^*. For an index 1 solution we have to give an appropriate row 0. The scheme (2.7) is the solution for K_7. For all values of the form $n = 12s + 7$ we will now construct row 0.

The vertices of K_n will be identified by the elements of the cyclic group $\mathbb{Z}_n$. We have to construct an appropriate row 0. We illustrate the method when $n = 19$. Row 0 has to be a permutation of $1, 2, \ldots, 18$ or $\pm 1, \pm 2, \ldots, \pm 9$.

Consider the graph in Fig. 2.13. The two ends marked A are to be identified. We then interpret A as an inner point of an arc and not a

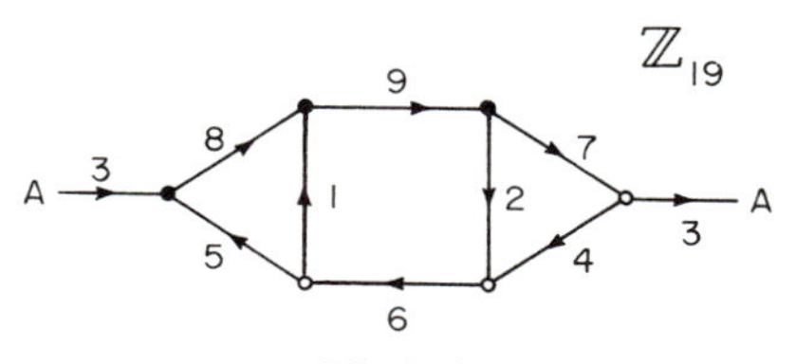

Fig. 2.13

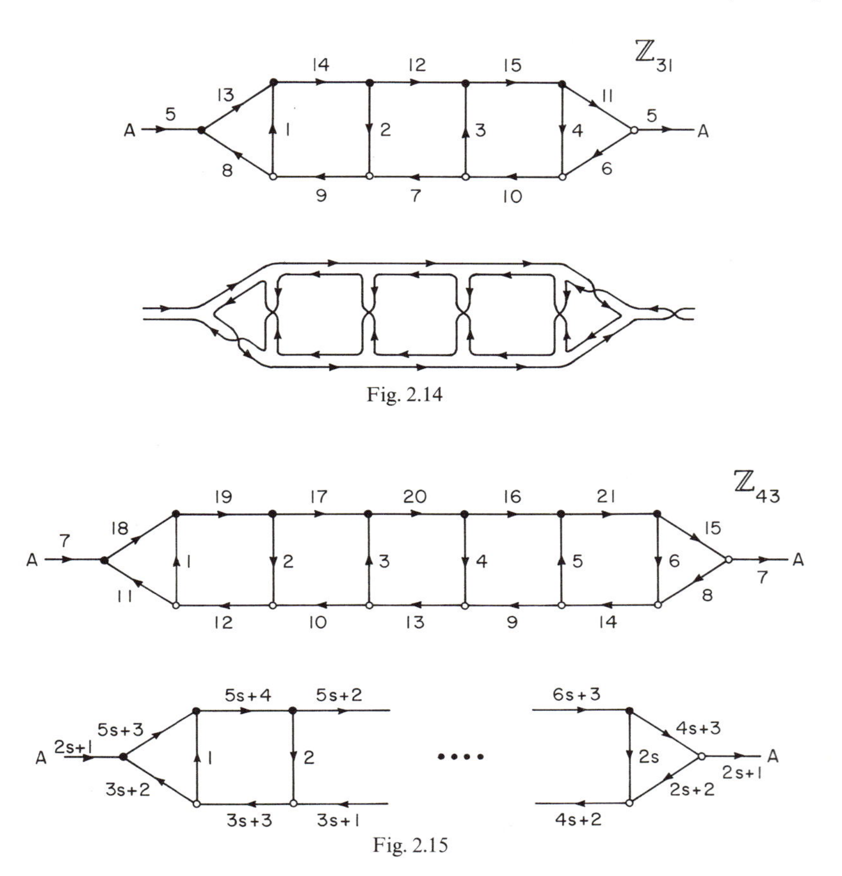

Fig. 2.14

Fig. 2.15

vertex. The same is to be done in Figs. 2.14, 2.15, 2.17. The graph of Fig. 2.13 is actually the same as G_5 in Fig. 2.1. Each arc in the graph of Fig. 2.13 has an orientation and a value called its current. The currents are elements of $\mathbb{Z}_{19}$. Notice that the graph in Fig. 2.13 (and respectively in Figs. 2.14, 2.15, 2.17, 2.18) has the following properties: (Later we will frequently use these properties as *construction principles* to produce schemes for graphs satisfying rule Δ^*.)

C 1) *Each vertex has valence* 3 *or* 1. (*Valence* 1 *will be used in Case* 10.)

C 2) *The given rotation is circular (induces one single circuit).* Recall • indicates clockwise and o counterclockwise rotation.

C 3) *Each element* $1, 2, \ldots, 6s+3$ *of* $\mathbb{Z}_{12s+7}$ (*here with* $s=1$) *appears exactly once as a current on some arc.*

The most important property is

C 4) *At each vertex of valence* 3 *the sum of the inward flowing currents equals the sum of the outward flowing currents.* This property is known as Kirchhoff's Current Law (shortened to KCL).

Actually, the graph in Fig. 2.13 is used as a kind of nomogram. It enables us to construct a triangular rotation of K_{19} in the following way. Consider the circuit which is induced by the rotation of the graph in Fig. 2.13. Record the currents sequentially, as they occur in the circuit, but if the direction of the circuit is opposite to the direction of the considered arc, record it with a minus sign. We will describe this in detail. It does not matter where we begin.

We start at the arc carrying current 9 in the direction indicated by the arrow. The next vertex is rotated clockwise so the next arc carries 7 and we are again in the direction of the arrow. We then get to a vertex which is rotated counterclockwise and continue on the arc with current 4. So far, we have recorded 9, 7, 4. The vertex on the end of arc 4 is again counterclockwise so we have to continue on the arc 2 but this time in the direction opposite that of the arrow. So we record $9, 7, 4, -2, \ldots$. The full *log* of the circuit is

$$9\,7\,4\,\bar{2}\,\bar{9}\,\bar{1}\,5\,\bar{3}\,\bar{7}\,2\,6\,1\,\bar{8}\,\bar{5}\,\bar{6}\,\bar{4}\,3\,8. \tag{2.10}$$

(It is convenient to omit the commas and write $\bar{a}$ instead of $-a$.) In the group $\mathbb{Z}_{19}$ the element -2 is the same as 17. Recording the elements in (2.10) in the positive form we obtain

$$9\ 7\ 4\ 17\ 10\ 18\ 5\ 16\ 12\ 2\ 6\ 1\ 11\ 14\ 13\ 15\ 3\ 8.$$

We take this as row 0 of a scheme and obtain row i by adding i to the terms in row 0 without changing the order ($i = 1, 2, \ldots, 18$).

We have to prove that this scheme satisfies rule Δ^*. Let us assume that in the scheme one finds

$$i. \ldots j\,k \ldots.$$

The additive rule says that

$$0. \ldots j-i\ k-i \ldots$$

appears in row 0. The local picture of the current graph is shown in Fig. 2.16, where h is the current of the third arc. (If some of the arrows

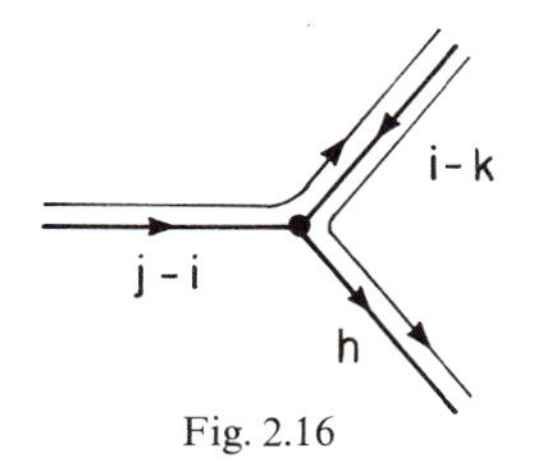

Fig. 2.16

in the picture point in the other direction we shall get the same result.) It follows that

$$0. \ldots j-i\; k-i \ldots i-k\; h \ldots .$$

Adding k we get row k as follows

$$k. \ldots i\; h+k \ldots .$$

Because of Kirchhoff's Current Law (property C4), $h=(j-i)+(i-k)$ and consequently $j=k+h$. Hence we have

$$k. \ldots i\, j \ldots$$

and rule Δ^* holds. It is shown that we really have constructed a triangular rotation of K_{19}.

The generalization for K_{12s+7} is easy. Take a ladder-like graph with $2s$ rungs as shown in Fig. 2.15. (For $s=2$ see Fig. 2.14.) Observe that the chosen rotation is circular. The $2s$ rungs are directed in an alternating fashion up and down and carry the currents $1, 2, 3, \ldots, 2s$. The lower horizontals including the two "triangular sides" point to the left and carry the currents

$$3s+2, 3s+3, \ldots, 4s, 2s+4, 4s+1, 2s+3, 4s+2, 2s+2.$$

This row consists of two dove-tailed arithmetic sequences with differences -1 and 1 respectively. Each of the upper horizontals points to the right and we get the appropriate current by adding $2s+1$ to the current of the corresponding lower horizontal. See Fig. 2.15. It is easy to check that the four properties C1) to C4) are valid. Figs. 2.14 ($s=2$) and 2.15 ($s=3$) show two more examples. Notice that one can obtain the rotation (2.7) of K_7 using Fig. 2.17.

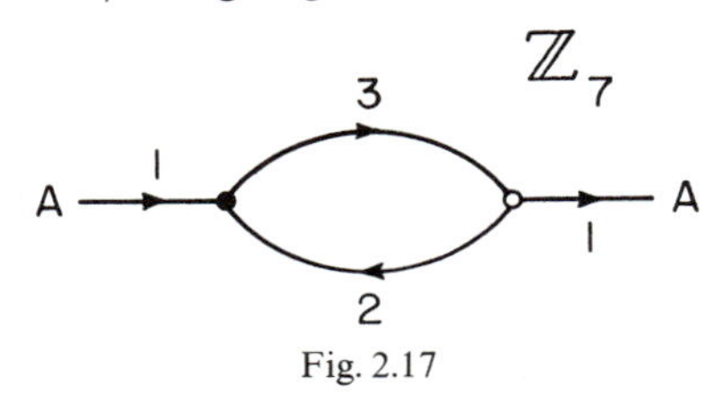

Fig. 2.17

The schemes (2.7) and (2.8) satisfy the

Rule R^*. *If in row i one has $i. \ldots j\,k\,l \ldots$ then row k appears as $k. \ldots l\,i\,j \ldots$.*

It is obvious that if a scheme for a graph G satisfies rule R^* then it satisfies rule Δ^*. We shall prove the converse statement. Suppose we have

$$i. \ldots j\,k\,l \ldots .$$

Using rule Δ^* one gets

$$k. \ldots i\,j \ldots \quad \text{and} \quad l. \ldots i\,k \ldots .$$

Apply rule Δ^* to row l to get

$$k. \ldots l\,i \ldots .$$

Since the graph G has no multiple arcs the symbol i appears exactly once in row k. Hence we have

$$k. \ldots l\,i\,j \ldots$$

and rule R^* is proved.

For checking an explicitly given scheme it is better to use rule R^* while rule Δ^* is more convenient for theoretical proofs.

Theorem 2.9. *For each positive integer* $n \equiv 10 \pmod{12}$ *there exists a triangular rotation of the graph* $K_n - K_3$.

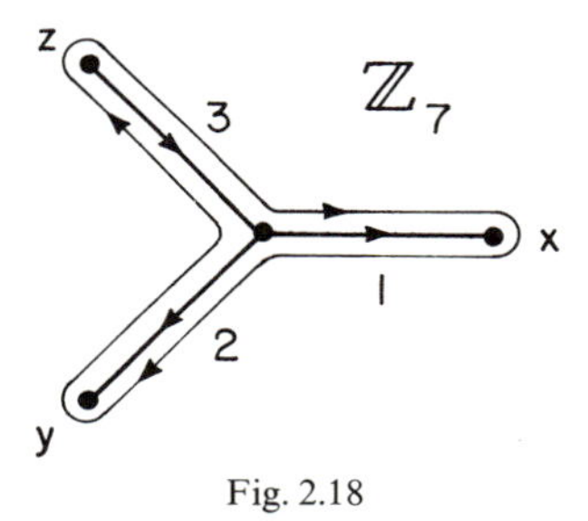

Fig. 2.18

In order to prove this we have to generalize the rotation (2.8) of $K_{10} - K_3$. This will be done by a slight modification of the current graph of Fig. 2.15. Let n be of the form $n = 12s + 10$. For $s = 0$ consider the current graph of Fig. 2.18. The four properties C1), C2), C3), C4) are satisfied. In regard to the vertices of valence 1, observe the new property

C5) *Each vertex identified by a letter, such as x, is of valence* 1 *and the current flowing into the vertex generates the group. This means that the greatest common divisor of this current and* $12s+7$ *is* 1.

A vertex where Kirchhoff's Current Law does not hold is called a *vortex*. So the vertices x, y, z are vortices.

When we describe the circuit of the current graph as in the previous case we have to write the letter x when we pass the vertex x. So we obtain the sequence

$$1\, x\, \bar{1}\, 2\, y\, \bar{2}\, \bar{3}\, z\, 3$$

from the current graph of Fig. 2.18. Taking this as row 0 and manufacturing rows $1, 2, \ldots, 6$ using the additive rule, and rows x, y, z using rule Δ^*, we obtain the scheme (2.8).

In the general case $n=12s+7$ modify the current graph of Fig. 2.15 by replacing the vertex at the left by three vortices x, y, z to get Fig. 2.19. Also change the counterclockwise rotation of the right hand vertex in order to get a circular rotation.

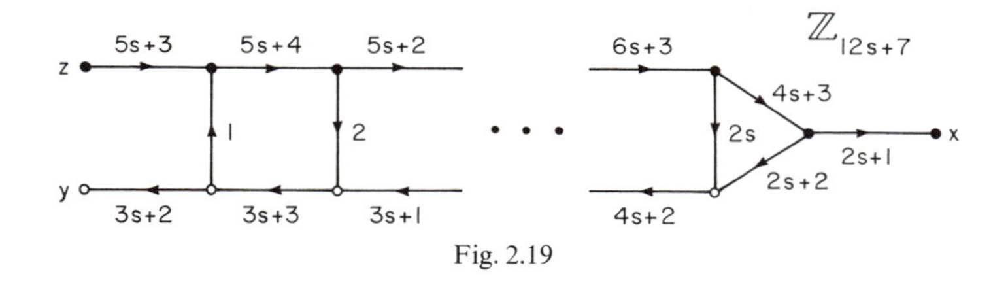

Fig. 2.19

If a, b are two integers the greatest common divisor is denoted by (a, b). Using the rule $(a, b)=(a, b-na)$ for any integer n one can calculate the following greatest common divisors.

$$(2s+1, 12s+7)=(2s+1, 1)=1,$$

$$(3s+2, 12s+7)=(3s+2, -1)=1,$$

$$(5s+3, 12s+7)=(5s+3, 2s+1)=(s+1, 2s+1)$$
$$=(s+1, -1)=1.$$

And so, the current on each of the three end arcs generates the group $\mathbb{Z}_{12s+7}$.

Later on we will often use such a vortex construction. Therefore it is worthwhile studying it in greater generality. To do so let a be the current on an end arc incident with the vortex x. The produced row 0 will have the part

$$0. \ldots a \;\; x \;\; \bar{a} \ldots$$

where $\bar{a}$ stands for $-a$. We therefore have

$$a. \ldots 2a \;\; x \;\; 0 \ldots$$

$$2a. \ldots 3a \;\; x \;\; a \ldots$$

in the scheme. Using rule R^* one gets for row x

$$x. \ldots \bar{a} \;\; 0 \;\; a \ldots.$$

Applying rule R^* repeatedly we get

$$x. \ldots \bar{a}\ 0\ a\ 2a\ 3a \ldots.$$

If a is a generator of the group, this row x contains all elements of the group.

Using the current graph of Fig. 2.19 we obtain a scheme for the graph $K_{12s+10}-K_3$. We just have to prove rule Δ^* in the following special case. Assume

$$i. \ldots x\ a+i \ldots.$$

If follows that

$$0. \ldots \bar{a}\ x\ a \ldots$$

and

$$a+i. \ldots i\ x \ldots.$$

In all other cases rule Δ^* is already proved. We have therefore constructed a triangular rotation for $K_{12s+10}-K_3$.

Later we will use the concept of a "global" Kirchhoff's Current Law. In the current graph of Fig. 2.13, consider the upper horizontal arc. The sum of all currents flowing into the *two* vertices incident with this arc is $8+1+9$, and the sum of all currents flowing out of the two vertices is $7+2+9$. These sums are equal, and in fact, if we cancel the two 9's we see that the sum of the currents flowing into this part of the graph $(8+1)$ equals the sum of those flowing out $(7+2)$.

Theorem 2.10 (Kirchhoff's Global Current Law). *Suppose G is a current graph satisfying KCL at each vertex, and H is a subgraph of G. We say an arc of G flows into (out of) H if the arc is incident with exactly one vertex P of H, and the direction of the arc is towards (away from) P. Then the sum of currents on arcs which flow into H equals the sum of currents on arcs which flow out of H.*

This theorem is easily proved by induction on the number of arcs in H.

In the example above, H was the subgraph consisting of the upper horizontal arc with its two vertices.

Exercises

In order to understand some of the following problems observe that in Fig. 2.13 KCL holds not only in the group $\mathbb{Z}_{19}$. It really holds in $\mathbb{Z}$. ($\mathbb{Z}$ is the additive group of all integers.) Also in all the other current graphs in this chapter KCL holds in $\mathbb{Z}$.

2.3.1. Multiply each current in Fig. 2.13 by 5 and reduce mod 19. We obtain a new current graph still satisfying KCL in $\mathbb{Z}_{19}$. More generally, if $(c, 12s+7)=1$ then by multiplying each current by c we get a new current graph in Figs. 2.13 and 2.15 satisfying KCL. Check this and try to prove it.

2.3.2. Distribute the elements 1, 2, 3, 4, 5, 6 of the group $\mathbb{Z}_{13}$ and put arrows on the six arcs of K_4 such that KCL holds in $\mathbb{Z}_{13}$. Why does this not lead to a triangular rotation of K_{13}?

2.3.3. Determine the number of triangles induced by a triangular rotation of K_{12s+7} and $K_{12s+10}-K_3$.

2.3.4. Consider the ladder-like graph as in Figs. 2.13, 2.14, 2.15 but with 5 rungs. Make a current graph with the group $\mathbb{Z}_{37}$ satisfying the construction principles C1), C3), and C4).

2.3.5. Find a triangular rotation for the graph $K_{15}-K_4$.

2.3.6. Find a triangular rotation for the graph $K_{25}-K_6$.

2.3.7. Find a triangular rotation for K_5-K_2, but use a current graph this time.

2.3.8. In generalization of problems 2.3.5 and 2.3.6 prove the theorem: If $8s+3$ is a prime number, then there exists a triangular rotation of the graph $K_{10s+5}-K_{2s+2}$.

2.3.9. Display the numbers $1, 2, \ldots, 12$ on the 12 arcs of a cube and put an arrow on each arc such that KCL holds in $\mathbb{Z}$ at each vertex.

2.3.10. Using the graph G_6 in Fig. 2.1 put arrows and currents of $\mathbb{Z}_{19}$ on the arcs and choose a rotation. Do this in such a way that the current graph produces a triangular rotation of K_{19}.

2.3.11. Show that the following is impossible to do: Put arrows on the arcs of the graph G_6 of Fig. 2.1 and display the numbers $1, 2, \ldots, 9$ on the arcs such that at each vertex KCL holds in $\mathbb{Z}$. (This time not only in $\mathbb{Z}_{19}$.)

2.3.12. If you display the numbers $2, 3, \ldots, 10$ instead of $1, 2, \ldots, 9$ then the construction in problem 2.3.11 is possible. Do it.

2.3.13. Show the unexpected fact that there is no triangular rotation of K_6-K_3.

2.3.14. Assign to each arc of the graph K_5 an arrow and one of the elements $1, 2, \ldots, 10$ such that KCL holds in $\mathbb{Z}$ and each of the numbers appears exactly once.

2.3.15. Show that it is not possible to use a current graph as in Fig. 2.20 (or more generally when the central part is a ladder with $2s-1$ rungs) with currents of $\mathbb{Z}_{15}$ (in general $\mathbb{Z}_{12s+3}$) in order to produce a triangular rotation of $K_{17}-K_2$ ($K_{12s+5}-K_2$).

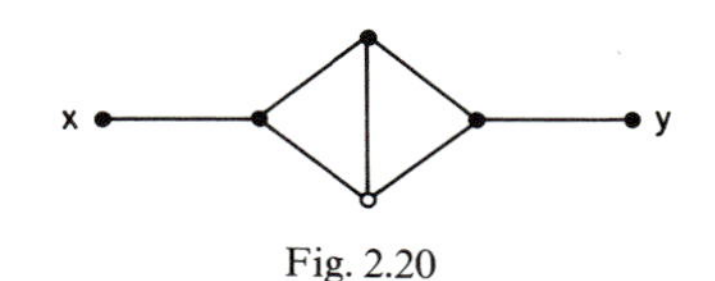

Fig. 2.20

2.3.16. Consider the currents in Fig. 2.13 as elements of the group $\mathbb{Z}_{20}$. The circular rotation of the graph defines a row 0 and by the additive rule one gets a scheme for a triangular rotation of a graph. Describe this graph.

3. Classification of Surfaces

3.1. The Concept of Topology

In this chapter we will present material that is well known; however we need not only the results, but also parts of the proofs as we shall see in later chapters. For more details of this theory see H. Seifert and Threlfall [81] or M. Fréchet and Ky Fan [22].

Suppose we are given a surface, made of an elastic material, e.g., India rubber. If we distort it in a continuous manner without tearing it, the resulting surface certainly will have the same chromatic number. Similarly if a graph which we may imagine to be made of wire is distorted without tearing, all of the properties of the graph considered in Chapter 2 remain unchanged: the valences of the vertices, the chromatic number, the connectivity, the length of the closed ways in the graph, and so on.

The study of those properties of a geometric figure which do not change after such a deformation is called *topology*. But it is necessary to give a precise mathematical definition for such a kind of deformation. This type of deformation that we are referring to is called a *homeomorphism*.

Although we will now define this concept rigorously, it is not necessary that the reader understand it in order to read the material following the definition.

Let E and F be point sets in Euclidean 3-space (or more generally in a metric space). Suppose that for each point x of E there exists a corresponding point $f(x)$ of F and for each point $y \in F$ there exists exactly one point x of E such that $f(x)=y$.

Such a *correspondence* or *mapping* f of E onto F is called a 1-1-*correspondence*. The inverse transformation which leads from each point y of F back to the original point x of E with $f(x)=y$ is normally denoted by f^{-1}, therefore we have $f^{-1}(y)=x$. Of course f^{-1} is a 1-1-correspondence of F onto E.

Example. Let E be a map of San Francisco and F San Francisco itself. Let f be the transformation which maps each point x of E exactly to that point in F which is represented by x. Then f is a 1-1-correspondence.

A transformation f of E onto F is defined to be *continuous* at a point x_0 of E, if for each number $\varepsilon>0$ there exists a number $\delta>0$ such that for every point x of E with a distance less than δ from x_0, the distance between the two points $f(x)$ and $f(x_0)$ in F is less than ε. If f is continuous at each point in E we say f is a *continuous transformation* of E *onto* F. If f is a 1-1-correspondence of E onto F and f and f^{-1} are both continuous, then f is called a *homeomorphism* or a *topological transformation*. The transformation f in the San Francisco example above is for instance a homeomorphism. If there exists a homeomorphism of E onto F then E and F are said to be *homeomorphic* and F is said to be the *topological image* of E. We present a list of examples of pairs of homeomorphic figures:

A circular disk, a square;
the surface of a sphere, the surface of a tetrahedron;
the interior of a circle, the interior of a rectangle;
a set of n distinct points, another set of n distinct points;
the surface of the torus, the surface of a tea cup;
a solid torus (with interior), a tea cup (with interior);
the plane, the interior of a circle;
the edges and vertices of an octahedron, three circles in the plane having 6 different intersecting points.

A circular disk is the surface (interior and the circumference) of a circle. Each topological image (considered in three dimensional space) of a circular disk is called a *country* or a 2-*cell*. Similarly the topological image of an edge is called an *arc* or a *simple curve* or a 1-*cell*. Consequently a 0-*cell* is just a point.

By the term *polygon* (triangle, quadrilateral, pentagon, hexagon, ...) we do not restrict ourselves to plane rectilinear figures. We define a *polygon with r sides* or an *r-gon* as a 2-cell which has its circumference divided into r arcs by r vertices. The arcs are the sides of the polygon. We also include the two cases $r=1$ and 2. A 2-gon is sometimes called a *lune*.

3.2. Polyhedra

We thus understand the term polygon in the sense of topology. Analogously we now define the term *polyhedron*. Before we present a general definition, let us consider some examples.

Examples of polyhedra: tetrahedron (Fig. 3.1), cube, octahedron, dodecahedron, icosahedron (these can all be considered as subdivisions of a sphere), the subdivisions of the torus into 7 hexagons as in Fig. 1.3, into 14 triangles as in Fig. 1.7, or into 16 quadrilaterals as in Fig. 3.2. The division of the torus into just one quadrilateral as in Fig. 1.2 will also be considered as a polyhedron.

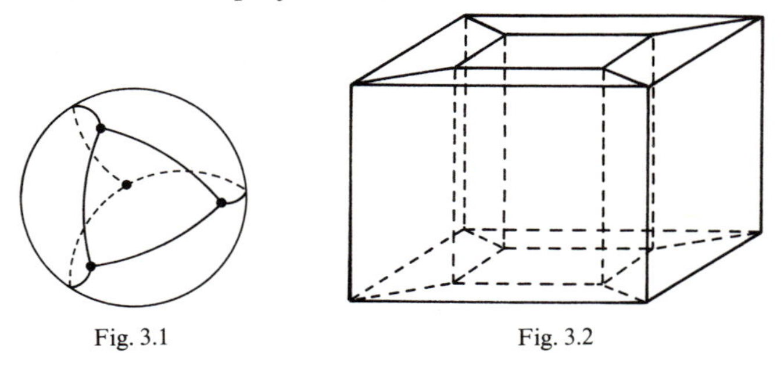

Fig. 3.1 Fig. 3.2

Fig. 1.2 shows how to divide the torus into one quadrilateral. We are now more interested in the reverse operation. Consider a rectangle (or quadrangle). Assign a definite sense of direction (orientation) to the perimeter of the rectangle which can be indicated by a curved arrow in the middle of the rectangle. Denote the sides by a, b, a, b in that order. Place an arrow on each side such that the two sides a (or b) are made to coincide and such that the heads of the arrows coincide as well. For this polyhedron we consider the "row"

$$a\,b\,a^{-1}\,b^{-1}. \tag{3.1}$$

as a symbolic description. It presents the four sides of the quadrangle. The notation a^{-1} in (3.1) means that the third side is denoted by a but its arrow points opposite to the given orientation of the quadrilateral.

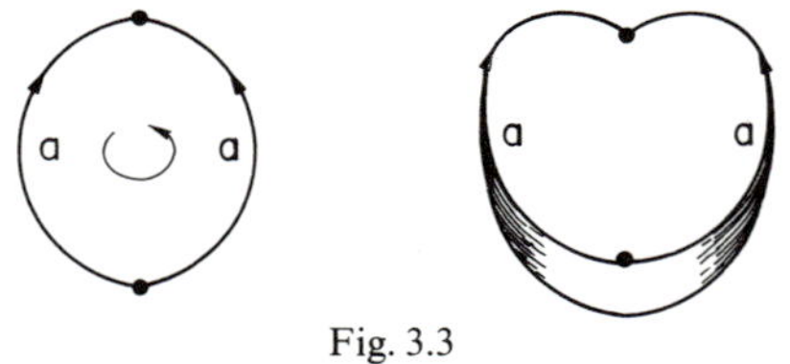

Fig. 3.3

As another example consider a lune (Fig. 3.3); it is a polygon with only two sides. Put arrows on the two sides and on the lune such that the symbolic row reads

$$a\,a^{-1} \tag{3.2}$$

This defines a division of the sphere into just one lune. It looks like a change purse where the two metal edges snap together.

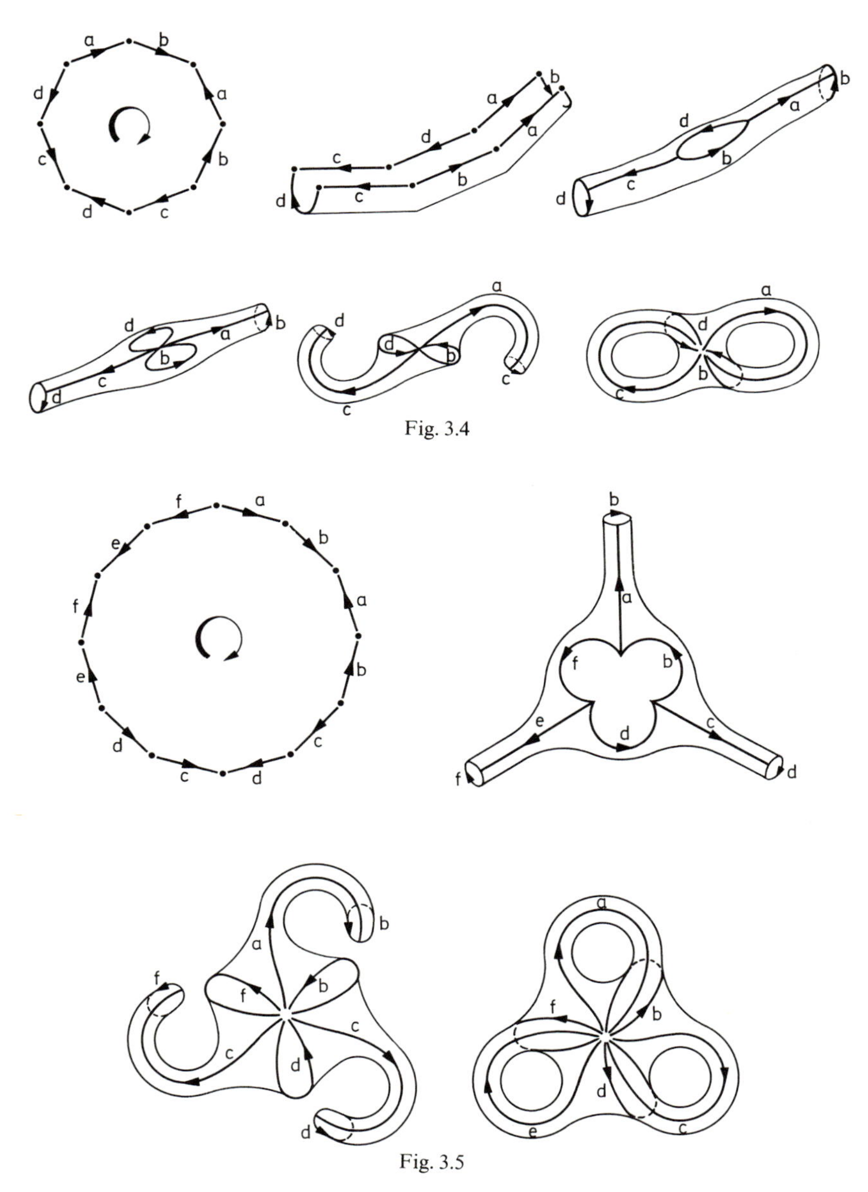

Fig. 3.4

Fig. 3.5

The following two examples are now easy to understand. The excellent Figs. 3.4 and 3.5 are taken from Hilbert and Cohn-Vossen [38]. Fig. 3.4 shows how to transform an octagon into a surface S_2. The corresponding symbolic notation is

$$(3.3) \qquad a b a^{-1} b^{-1} c d c^{-1} d^{-1}.$$

Similarly the 12-gon, as shown in Fig. 3.5, with the symbolic notation

$$(3.4) \qquad a b a^{-1} b^{-1} c d c^{-1} d^{-1} e f e^{-1} f^{-1}$$

describes a surface which we have already denoted by S_3. This can obviously be generalized for the sphere with p handles, S_p. Later we will return to this example.

If the polyhedron has s polygons, then the symbolic representation has s rows. For instance one can easily check that the scheme

$$\begin{matrix} e & d & f \\ a & c^{-1} & e^{-1} \\ f^{-1} & b & a^{-1} \\ c & b^{-1} & d^{-1} \end{matrix}$$

represents the tetrahedron as a subdivision of the sphere (see Fig. 3.6).

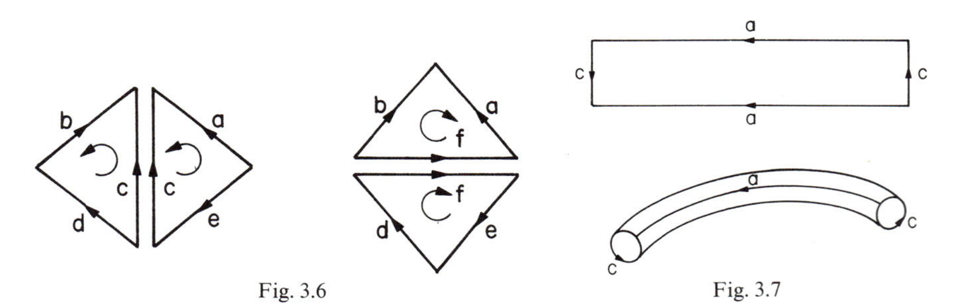

Fig. 3.6 Fig. 3.7

There are also polyhedra which cannot be embedded into three dimensional space without penetrating themselves. Consider a rectangle with the symbolic row

$$(3.5) \qquad c a c a^{-1}$$

as in Fig. 3.7. If we first make the identification along the sides labeled c we clearly get a Möbius strip; therefore the surface is going to be unilateral. It is better, however, in order to picture this geometrically, to first identify the sides labeled a. We obtain a tube for which the two end arcs are denoted by c. But to identify the two end circles, it is necessary to match the arrows on them. Deform the tube so that one of the ends is a little smaller than the other and penetrate this smaller end through

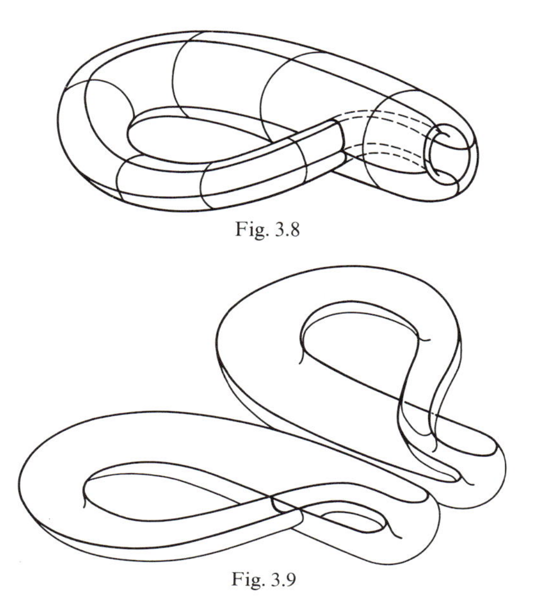

Fig. 3.8

Fig. 3.9

the wall of the tube. Then bend the larger end a little toward the interior and the smaller end a little toward the exterior, and attach them according to the required mode of identification (see Figs. 3.8 and 3.9). This surface is called a *Klein's bottle.*

We still have not given a precise definition for a polyhedron. We proceed as follows:

Given a finite number of polygons, let the total number of all of the sides of the polygons be even. Suppose these sides are given in pairs. Label the sides by letters such that two sides of a pair get the same letter. (Compare with Fig. 3.6.) Furthermore suppose that for each side an orientation is given and indicated by an arrow. Now identify each pair of sides such that the heads of the two arrows coincide. Two superimposed sides are called an *edge.* If the pseudograph consisting of the vertices and edges is connected, the figure so obtained is called a *polyhedron.* The polygons constituting a polyhedron are called the *faces* of the polyhedron.

All the information – the polygons, the pairing of the sides, and the orientation of the sides (i.e., before identification) – is called the *plane representation of the polyhedron.* If the total number of all of the sides of the polygons is more than two, many possibilities exist for matching the sides in pairs and for choosing an orientation for each side. Every possibility is allowed provided that the figure obtained is connected. For instance, one may pair two sides of the same polygon.

We can get a symbolic description of a polyhedron in the following way: Consider the plane representation of the polyhedron (i.e., before identification). For each polygon choose an orientation. Then, for each polygon, according to its chosen orientation write down the cyclic order of the sides as a row of the corresponding letters, and put the exponent -1 on the letter if the orientation of the polygon is opposite to the orientation of the side.

We obtain a scheme

$$(\Sigma) \qquad \begin{matrix} a \quad b \quad c^{-1} \quad \dots \\ \dots\dots\dots\dots\dots\dots \\ \dots\dots\dots\dots\dots\dots \end{matrix}$$

with the following properties:

α) Each letter appears exactly twice in (Σ).

β) The set of the rows in (Σ) cannot be separated into two disjoint subsets such that property α) holds for each of these subsets.

The property β) guarantees that the closed surface will be connected. It is obvious that the rows in (Σ) have to be read cyclically. The following operations change the scheme (Σ) but not the presented polyhedron.

a) Put the first letter of a row behind the last one (make a cyclic permutation).

b) Whenever a certain letter appears, say b (this will occur exactly twice, the second time, again as b or as b^{-1}), replace b by b^{-1} and b^{-1} by b. (We say we change the exponents.)

c) Do both of the following operations simultaneously: Change all the exponents in a row and reverse the cyclic order in this row.

The operations b) or c) can be interpreted as the reversing of the orientation of an edge or of a polygon respectively.

3.3. Elementary Operations

We wish to get a general idea of all possible kinds of polyhedra. We proceed as follows: Given a polyhedron P and its symbolic representation by the scheme (Σ), we shall define four operations which we call elementary operations. These operations do change the polyhedron as well as the representative scheme. But we know intuitively that the "surface" itself is not changed.

SU 1) *Subdivision of dimension one.*

An edge a of the polyhedron is divided into two new edges b, c by taking an inner point of a as an additional vertex (Fig. 3.10). In the scheme one has to replace the letter a by $b\,c$ and a^{-1} by $c^{-1}\,b^{-1}$.

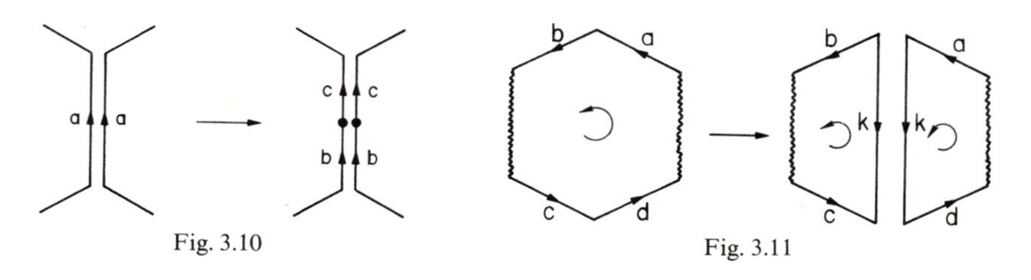

Fig. 3.10

Fig. 3.11

CO 1) *Composition of dimension one.*

This is the reverse operation of SU 1). If, in the scheme, we see two successive letters $a b$ in two places, replace $a b$ by c in both places. If we see $a b$ and $b^{-1} a^{-1}$ in the scheme replace these by c and c^{-1}.

SU 2) *Subdivision of dimension two.*

Two vertices of a polygon in the polyhedron will be connected by a new edge k dividing the polygon into two new polygons (Fig. 3.11). Instead of the single row

$$b \ldots c\, d \ldots a \tag{3.6}$$

in the scheme we write the two new rows

$$\begin{gathered} b \ldots c\, k^{-1} \\ k\, d \ldots a . \end{gathered} \tag{3.7}$$

CO 2) *Composition of dimension two.*

This is the inverse operation of SU 2). Two polygons with a common edge will be turned into one new polygon by omitting this common edge. In the scheme replace two rows of the form (3.7) by just one row of the form (3.6).

Two polyhedra P and P' are defined to be *elementarily related* if P can be transformed into P' by using a finite number of the elementary operations SU 1), CO 1), SU 2), CO 2).

We call a polyhedron *orientable* if one can choose an orientation for each polygon such that each edge is used in both possible directions. For the symbolic representation, this means the following: A polyhedron is orientable if, by applying operation c) as often as necessary, a scheme can be obtained in which for each side a, both a and a^{-1} appear in the symbolic representation.

It is easy to see that each of the operations SU 1), CO 1), SU 2), CO 2) transforms any orientable polyhedron again into an orientable polyhedron and therefore any nonorientable one into a nonorientable one.

In a polyhedron we denote the number of vertices, edges and polygons by α_0, α_1, and α_2 respectively. (These numbers are to be understood of course *after* identifying all the equivalent pairs of sides of polygons.) For instance, in the subdivision of the torus into one quadrilateral as shown in Fig. 1.2, the numbers are: $\alpha_0=1$, $\alpha_1=2$, $\alpha_2=1$. In the symbolic representation of a polyhedron, α_2 is simply the number of rows and α_1 is the number of different letters. The determination of α_0 is a little more complicated. We do not consider this now.

The alternating sum

$$E(P)=\alpha_0-\alpha_1+\alpha_2 \tag{3.8}$$

for a polyhedron P is called its *Euler characteristic*. It does not change when applying one of the elementary operations. In the case of SU 1) the numbers α_0 and α_1 both increase by one, and therefore $E(P)$ stays constant. In the case of SU 2) the numbers α_1 and α_2 both increase by one, and consequently $E(P)$ is unchanged. The operations CO 1) and CO 2) do not change the Euler characteristic because their inverse operations SU 1) and SU 2) do not.

Summarizing this we can easily see that the following statement is true:

Theorem 3.1. *Two elementarily related polyhedra have the same Euler characteristic and are both orientable or both nonorientable.*

Our next aim is to prove that the converse statement of Theorem 3.1 is also true. So we shall prove that two orientable (and also two nonorientable) polyhedra with the same Euler characteristic are elementarily related.

The set of all polyhedra is divided into pairwise disjoint classes in such a way that two polyhedra belong to the same class if and only if they are elementarily related. In each class we are going to find the simplest polyhedron which we shall call the "normal form" of the class. Of course we have to define what we really mean by "simplest polyhedron" in a class. In constructing the normal form for a given polyhedron the symbolic representation will be of great help.

3.4. Normal Form for Orientable Surfaces

We consider a given orientable polyhedron and are going to apply the following reductions.

1. *Producing One Row.* If the polyhedron contains more than one polygon, we can use the composition of dimension two a finite number of times such that we finally get a polyhedron with only one polygon. The symbolic representation has just one row.

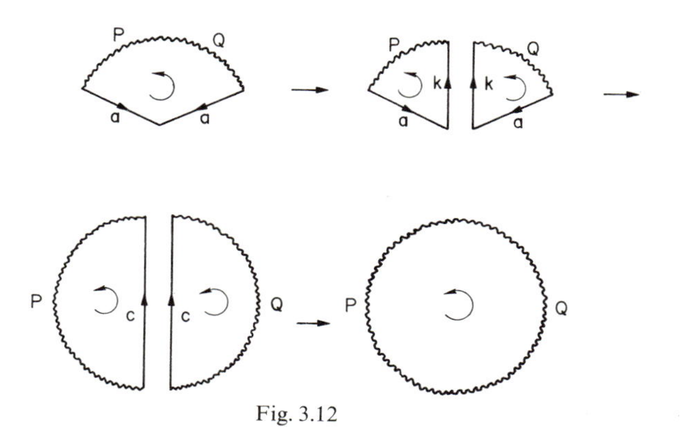

Fig. 3.12

2. *Simple Normalization.* It is convenient to denote a sequence of small letters in a row by one capital letter. So the row $P\,a\,a^{-1}\,Q$ is a short notation for a row of the form $b_1\,b_2\,\dots\,b_s\,a\,a^{-1}\,c_1\,c_2\,\dots\,c_t$.

Let $P\,a\,a^{-1}\,Q$ be a row in the symbolic representation of a polyhedron and let P and Q both be not empty. If we replace $P\,a\,a^{-1}\,Q$ by PQ we get a representation for a polyhedron which is elementarily related to the first polyhedron. The proof is given by applying the operations SU 2), CO 1), CO 2):

$$P\,a\,a^{-1}\,Q \to \begin{cases} P & a & k \\ k^{-1} & a^{-1} & Q \end{cases} \to \begin{cases} P & c \\ c^{-1} & Q \end{cases} \to PQ.$$

Fig. 3.12 illustrates this proof. We call this transformation

$$P\,a\,a^{-1}\,Q \to PQ$$

a *simple normalization.*

After producing one row and then applying simple normalization as often as possible, we will arrive at one of two possible cases:

I. The polyhedron has only one edge and therefore it has the form $a\,a^{-1}$ exactly as in example (3.2) of Fig. 3.3. This we consider as a normal form. We say $a\,a^{-1}$ is the normal form of the sphere.

II. The polyhedron has more than one edge (this means in the row of the symbolic representation there are at least two different letters). In this case the process of reductions continues as follows.

3. *Preparation for Handle Normalization.* In the row representing the polyhedron there are two letters a, b such that they appear in the following (alternating) form

$$\dots\,a\,\dots\,b\,\dots\,a^{-1}\,\dots\,b^{-1}\,\dots\,.$$

In order to prove this consider a letter a in the given row. Then the row has the form $aQa^{-1}R$ where Q is not empty. Otherwise, simple normalization would be possible. Now choose the letter a such that Q is as short as possible. If b is a letter from Q then b^{-1} must appear in R because the sequence Q has minimal length.

4. *Handle Normalization.* We have shown that the row can be written in the form $PaQbRa^{-1}Sb^{-1}T$. Apply elementary operations several times:

$$PaQbRa^{-1}Sb^{-1}T \to \begin{cases} PaQbRc \\ c^{-1}a^{-1}Sb^{-1}T \end{cases} \to \begin{cases} QbRcPa \\ a^{-1}Sb^{-1}Tc^{-1} \end{cases}$$

$$\to QbRcPSb^{-1}Tc^{-1} \to c^{-1}QbRcPSb^{-1}T \to \begin{cases} c^{-1}QbRck \\ k^{-1}PSb^{-1}T \end{cases}$$

$$\to \begin{cases} Rc\,kc^{-1}Qb \\ b^{-1}Tk^{-1}PS \end{cases} \to Rckc^{-1}QTk^{-1}PS \to c^{-1}QTk^{-1}PSRck$$

$$\to \begin{cases} c^{-1}QTj \\ j^{-1}k^{-1}PSRck \end{cases} \to \begin{cases} QTjc^{-1} \\ ckj^{-1}k^{-1}PSR \end{cases} \to QTjkj^{-1}k^{-1}PSR$$

$$\to PSRQTjkj^{-1}k^{-1} \to PSRQTaba^{-1}b^{-1}.$$

In the last step we have just replaced the two letters j, k by a, b. This transformation

$$PaQbRa^{-1}Sb^{-1}T \to PSRQTaba^{-1}b^{-1}$$

is called a *handle normalization.* Fig. 3.13 tries to illustrate the above transformation: First divide the polygon by a new diagonal c into two polygons. They are pasted together along the edge a obtaining one polygon. This is divided by the new diagonal k. The two parts are pasted together along b. The polygon obtained is divided by j. The two parts are glued together along c.

5. *Constructing the Normal Form.* In the row $PSRQTa_1b_1a_1^{-1}b_1^{-1}$ we apply simple normalization as often as possible. If all the letters in $PSRQT$ are canceled then we consider the row $a_1b_1a_1^{-1}b_1^{-1}$ as a normal form. It is the normal form of the torus. If $PSRQT$ remains non-empty then, as in step 3 (Preparation for Handle Normalization, Section 3.4), the row can be written in the form

$$P_2a_2Q_2b_2R_2a_2^{-1}S_2b_2^{-1}T_2a_1b_1a_1^{-1}b_1^{-1}.$$

Applying again a handle normalization with $T = T_2a_1b_1a_1^{-1}b_1^{-1}$ we get

$$P_2S_2R_2Q_2T_2a_1b_1a_1^{-1}b_1^{-1}a_2b_2a_2^{-1}b_2^{-1}.$$

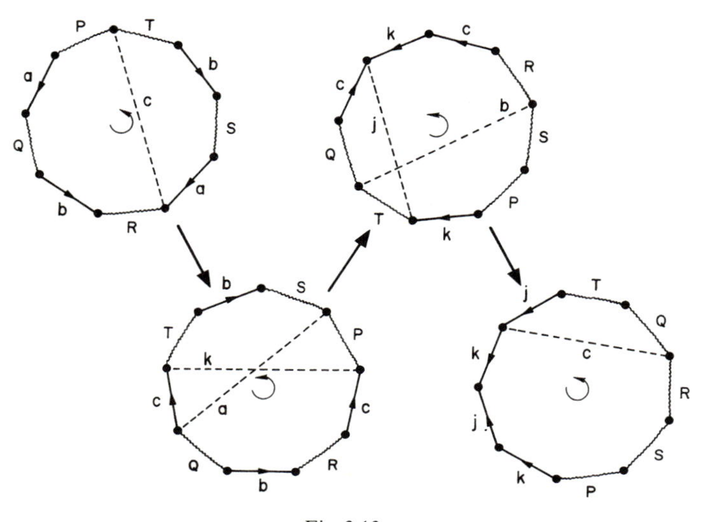

Fig. 3.13

Then reduce the sequence $P_2\, S_2\, T_2\, Q_2\, T_2$ by simple normalization. If some letters still remain we can again use handle normalization. Continuing in this manner we finally get a normal form

$$(H_p) \qquad a_1\, b_1\, a_1^{-1}\, b_1^{-1}\, a_2\, b_2\, a_2^{-1}\, b_2^{-1} \ldots a_p\, b_p\, a_p^{-1}\, b_p^{-1}.$$

In (3.3) and (3.4) we have already considered this normal form for $p=2$ and $p=3$ (Figs. 3.4 and 3.5). We are going to determine the Euler characteristic of the polyhedron (H_p). Observe that before we identify the sides, the polygon has $4p$ vertices.

Denote them by

$$P_1, P_2, \ldots, P_{4p},$$

where

$$a_1=(P_1, P_2), \quad b_1=(P_2, P_3), \quad a_1^{-1}=(P_3, P_4), \quad b_1^{-1}=(P_4, P_5), \quad a_2=(P_5, P_6),$$

and so on. By the process of identification it follows that $P_1=P_4$, $P_2=P_3$, $P_3=P_4$, $P_2=P_5$, ... Finally it follows that the polygon (H_p) really has *one* vertex. Set $\alpha_0=1$, $\alpha_1=2p$, $\alpha_2=1$ in formula (3.8) to get $E(H_p)=\alpha_0-\alpha_1+\alpha_2=2-2p$. Therefore the Euler characteristic of (H_p) is $2-2p$.

3.5. Normal Form for Non-Orientable Surfaces

Suppose we are given a non-orientable polyhedron and its symbolic representation. The following steps will lead to the normal form. First, use simple normalization as often as possible. Then apply composition of dimension two so as to obtain one single row (one polygon). If P is the notation for a sequence of letters, e.g. $P=abcd$, the notation P^{-1} means $P^{-1}=d^{-1}c^{-1}b^{-1}a^{-1}$ (reversing the order and changing all the exponents). The following steps will lead to the normal form.

Cross Cap Normalization. The symbolic representation of the non-orientable polyhedron is now one row of the form $PcQcR$. If P, Q, and R are empty then we consider the row cc as a normal form. If not, we use elementary operations as follows:

$$PcQcR \to \begin{cases} Pck^{-1} \\ kQcR \end{cases} \to \begin{cases} P^{-1}kc^{-1} \\ cRkQ \end{cases} \to P^{-1}kRkQ \to kRkQP^{-1}$$

$$\to \begin{cases} kRc \\ c^{-1}kQP^{-1} \end{cases} \to \begin{cases} Rck \\ k^{-1}cPQ^{-1} \end{cases} \to RccPQ^{-1} \to PQ^{-1}Rcc.$$

This transformation

$$PcQcR \to PQ^{-1}Rcc$$

is called *cross cap normalization.*

Transformation of One Handle Into Two Cross Caps. Assume the symbolic representation of the polyhedron is given in the form $Rccaba^{-1}b^{-1}$. Then we can use the following transformation.

$$Rccaba^{-1}b^{-1} \to a^{-1}b^{-1}Rccab \to \begin{cases} a^{-1}b^{-1}Rck \\ k^{-1}cab \end{cases}$$

$$\to \begin{cases} Rcka^{-1}b^{-1} \\ bk^{-1}ca \end{cases} \to Rcka^{-1}k^{-1}ca.$$

We now use cross cap normalization twice, first with respect to c, then with respect to a.

$$Rcka^{-1}k^{-1}ca \to Rkak^{-1}acc \to Rkkccaa \to Rccaabb.$$

In the last step we have changed some of the letters.

Constructing the Normal Form. Given a non-orientable polyhedron and applying composition of dimension 2 several times, we get one single row as the symbolic representation. Use simple normalization and then cross cap normalization as often as possible. We obtain a row of the form

$$W c_1 c_1 c_2 c_2 \dots c_t c_t.$$

In the sequence W each pair of letters appears with different exponents. Otherwise one more cross cap normalization could be used. Since W itself is orientable it can be written in the form

$$W = P_1 a_1 Q_1 b_1 R_1 a_1^{-1} S_1 b_1^{-1} T_1$$

exactly as it was in the orientable case. Applying handle normalization we obtain the row

$$W_1 c_1 c_1 c_2 c_2 \dots c_t c_t a_1 b_1 a_1^{-1} b_1^{-1}.$$

If W_1 is not empty after using simple normalization we can use handle normalization once more. When we repeat this process as often as possible we get the row

$$c_1 c_1 c_2 c_2 \dots c_t c_t a_1 b_1 a_1^{-1} b_1^{-1} \dots a_s b_s a_s^{-1} b_s^{-1}.$$

Since $t \geqq 1$ we can transform one handle after the other into cross caps. Finally we get the row

$$(C_q) \qquad c_1 c_1 c_2 c_2 \dots c_q c_q$$

which we call the *normal form* (C_q) of the given non-orientable polyhedron.

Label the vertices of the $2q$-gon (C_q) *before* identification of the edges by

$$P_1, P_1', P_2, P_2', \dots, P_q, P_q'$$

in this order such that P_1 is the vertex between the two sides both labeled by c_1.

Then identification of the sides shows us easily that $P_1 = P_1' = P_2 = P_2'$ and so on. Therefore the normal form (C_q) has one vertex, q edges, and one polygon. Hence the Euler characteristic of the polyhedron (C_q) is

$$E(C_q) = 2 - q.$$

3.6. Standard Models

We have found that each polyhedron is elementarily related to one of the following normal forms:

	Normal Form	Euler Characteristic	Orientable
(H_0)	$a a^{-1}$	2	yes $(p=0)$
(H_p)	$a_1 b_1 a_1^{-1} b_1^{-1} \dots a_p b_p a_p^{-1} b_p^{-1}$	$2-2p$	yes $(p=1, 2, \dots)$
(C_q)	$c_1 c_1 c_2 c_2 \dots c_q c_q$	$2-q$	no $(q=1, 2, \dots)$

The following two theorems are now obvious.

Theorem 3.2. *For each polyhedron P the Euler characteristic $E(P)$ is less than or equal to 2.*

Theorem 3.3. *The polyhedron P is elementarily related to the normal form $a a^{-1}$ (sphere) if and only if $E(P)=2$.*

Any two different normal forms are not elementarily related because they differ either in the Euler characteristic or in the kind of orientability.

Theorem 3.4. *Two polyhedra are elementarily related if and only if both have the same Euler characteristic and both are orientable or both are not orientable.*

Theorem 3.1 shows that the condition is necessary. In order to prove that it is also sufficient assume P and P' are two orientable polyhedra with $E(P)=E(P')$. Both polyhedra P and P' must be elementarily related to the same normal form, because there is only one orientable normal form with a given Euler characteristic. Therefore P is elementarily related to P'. The same argument can be used if P and P' are both non-orientable.

It is convenient at this stage of the theory to define a surface in the following combinatorial way.

If P is a polyhedron, we call the set S of all polyhedra elementarily related to P a *closed surface*. If P is a member of the set S we may also say P is a *polyhedron on* S or P is a subdivision of S.

The *Euler characteristic* $E(S)$ of a closed surface is defined by $E(S)=E(P)$ if P is a polyhedron on S. We also say that S is *orientable* (*non-orientable*) if P is orientable (non-orientable). This is well defined because each polyhedron on S has the same Euler characteristic and the same kind of orientability.

The set of all polyhedra elementarily related to the normal form (H_p) we call the *closed orientable surface of genus* p and denote this set by S_p $(p=0, 1, 2, \ldots)$.

Analogously, we call the set of all polyhedra elementarily related to the normal form (C_q) the *closed non-orientable surface* of genus q and denote it by N_q $(q=1, 2, \ldots)$.

The surface S_0 is normally called the *sphere* and S_1 is the *torus*. We are going to show that a sphere with p handles can be used as a *standard model* for the surface S_p. In the special cases $p=1, 2, 3$ this was already illustrated in Figs. 1.2, 3.4, and 3.5.

The Fig. 3.14 illustrates an orientable polyhedron consisting of 6 quadrilaterals and two 12-gons. It is easy to see that two quadrilaterals form one handle, and one of the 12-gons can be considered as the northern hemisphere, the other as the southern hemisphere. The arrows on the

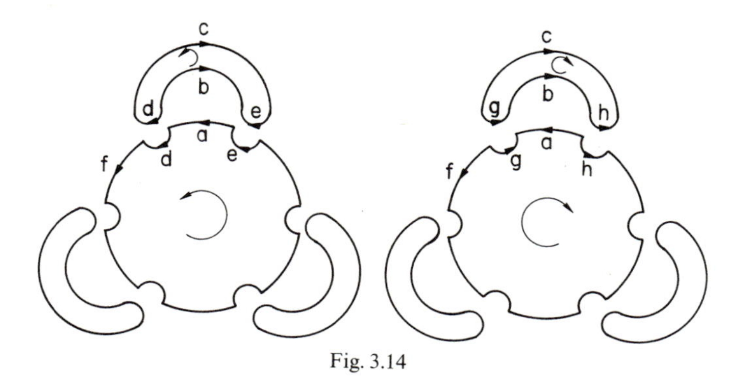

Fig. 3.14

polygons make evident that the polyhedron is orientable. The generalization for p handles is clear: It will have $2p$ quadrilaterals and two $4p$-gons. From the equations

$$\alpha_0 = 4p, \quad \alpha_1 = 8p, \quad \alpha_2 = 2p + 2$$

we obtain the Euler characteristic $2-2p$. Hence the sphere with p handles can be considered as a model for the orientable surface of genus p.

For a nonorientable surface there is not an easy model available. Consider the polyhedron consisting of one 12-gon and one hexagon with the plane representation as in Fig. 3.15. It has the Euler characteristic

$$\alpha_0 - \alpha_1 + \alpha_2 = 6 - 9 + 2 = 2 - 3$$

and is non-orientable. Hence it represents the non-orientable surface of genus 3. In general we consider the Möbius strip M_q of order q which we have described in Section 1.4. M_q has one single closed curve as boundary. We take an additional 2-cell which also has one single boundary curve. When we identify the two boundary curves we get a representation of the non-orientable surface of genus q.

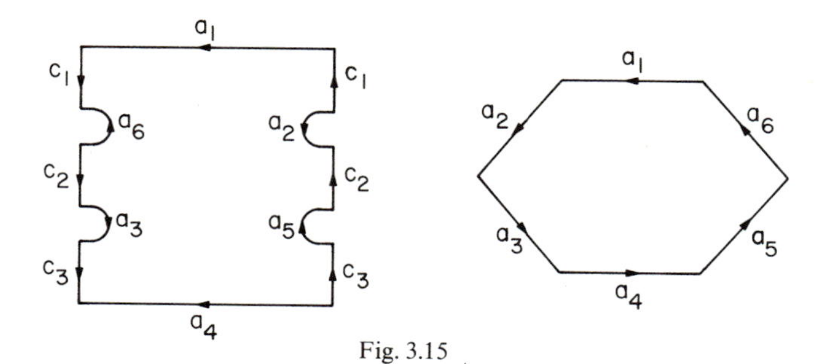

Fig. 3.15

If we cut one little 2-cell out of the surface N_q we get the surface M_q. M_q is easy to imagine geometrically, while it is almost impossible for N_q. In fact it is impossible to find an embedding of N_q into Euclidean three dimensional space without penetrating parts of N_q through itself. As we have already seen the surface N_2 is the same as Klein's bottle.

Exercises

3.6.1. Transform the polygon $abca^{-1}b^{-1}c^{-1}$ into one of the normal forms.

3.6.2. Transform the polygon $abca^{-1}b^{-1}c$ into one of the normal forms.

3.6.3. What surface does the following polygon represent:

$$a_1 a_2 \ldots a_n a_1^{-1} a_2^{-1} \ldots a_n^{-1}?$$

The answer can be given without transforming into normal form.

3.6.4. What is wrong in the following "proof" for the cross cap normalization?

$$PcQcR \to PcQP^{-1}PcR \to \begin{cases} PcQP^{-1}k^{-1} \\ kPcR \end{cases} \to \begin{cases} P^{-1}kPQ^{-1}c^{-1} \\ cRkP \end{cases}$$
$$\to P^{-1}kPQ^{-1}RkP \to kkPQ^{-1}R.$$

3.6.5. Given a polyhedron with the properties: Each face is a pentagon and each vertex is of valence four. The number of pentagons is greater than 8 and less than 16. Show that this polyhedron is non-orientable.

3.6.6. Given a polyhedron with the properties: Each face is a pentagon or a hexagon, each vertex is of valence three, and the number of pentagons is $\geqq 12$. Show that this polyhedron is orientable.

3.6.7. Assume each face of a polyhedron on the torus is a quadrilateral or a triangle and each vertex is of valence four or three. What is the number of triangles and the number of vertices of valence three?

3.6.8. What surface do we obtain if we change the direction of one of the 12 arrows in Fig. 3.6?

3.6.9. Given a $2n$-gon, there are many ways to match the sides in pairs and then each pair can be identified in two different ways.

a) If the polyhedron obtained is orientable, what is its highest possible and smallest possible genus?

b) If the polyhedron obtained is non-orientable, what is its highest possible and smallest possible genus? (The answer to a) depends on the parity of n (even or odd).)

3.6.10. Given $2n$ triangles, there are many ways to identify the sides in pairs in order to get one polyhedron. Find the answer to the same two questions a) and b) as in the previous exercise number 3.6.9.

3.6.11. Transform the polygon $a_1 a_2 \ldots a_n a_n \ldots a_2 a_1$ into the normal form.

3.6.12. How many elementary operations are needed to transform a cube into a tetrahedron?

3.7. Partial Polyhedra

Given a finite number of polygons, consider a *proper* subset D of the polygon sides. Let the number of sides in D be even. Label the sides in D by letters such that each of the letters appears on exactly two sides. Orient each of the sides in D by assigning an arrow. Then you get the so-called *plane representation* of a *partial polyhedron.* When we identify the sides labeled with the same letter such that the heads of the arrows coincide we get a *partial polyhedron.* The sides which do not belong to D are unlabeled and unoriented and we call them the *boundary edges* of the partial polyhedron. By definition there is at least one boundary edge.

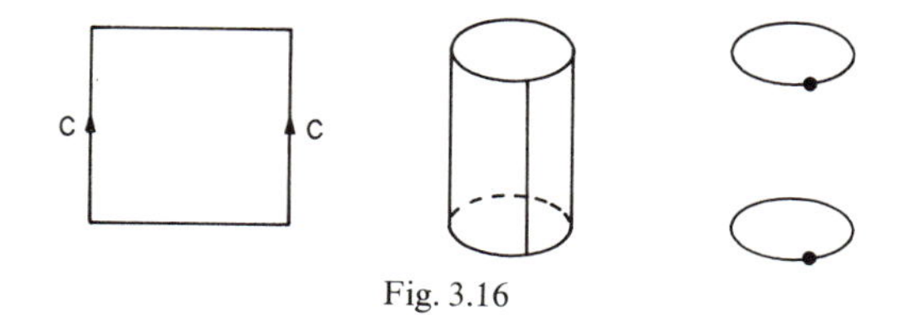

Fig. 3.16

In a partial polyhedron the boundary edges and the vertices which are incident with at least one boundary edge form a pseudograph. We call it the *boundary pseudograph* of the partial polyhedron. Fig. 3.16 illustrates the plane representation (left hand part) of a partial polyhedron (central part) and its boundary pseudograph (right part). Another example is given by Fig. 3.17. The Möbius strip (or more generally M_q from Section 1.4, Fig. 1.10) can be considered as a partial polyhedron.

Theorem 3.5. *Each vertex of the boundary pseudograph of a partial polyhedron is of valence* 2.

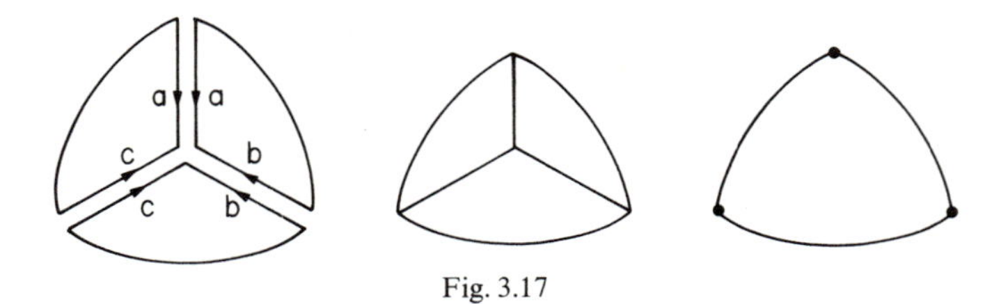

Fig. 3.17

Proof. Consider in the plane representation a polygon with a vertex A_0 incident with a boundary side (unlabeled side). If this vertex A_0 is incident with no labeled side of the polygon then val $A_0=2$ even after identifying all of the labeled sides. If not, let c_1 be the labeled side of the polygon incident with A_0 (see Fig. 3.18). There exists a sequence

$$A_0, c_1; c_1, A_1, c_2; c_2, A_2, c_3; \ldots; c_t, A_t$$

such that two successive letters c_i; c_i represent two different sides both labeled by c_i ($i=1, 2, \ldots, t$), and c_i, A_i, c_{i+1} belong to a polygon such that A_i is incident with c_i and c_{i+1} ($i=1, 2, \ldots, t-1$), and the arrows on the c_i are as follows: The first c_i points to A_{i-1} if and only if the second c_i points to A_i. If the above-mentioned sequence is as long as possible the last vertex A_t is incident with an unlabeled side. After identifying all the sides, all the vertices A_i ($i=0, 1, \ldots, t$) are identified as one vertex which will only be incident with exactly two unlabeled sides (or one as a loop).

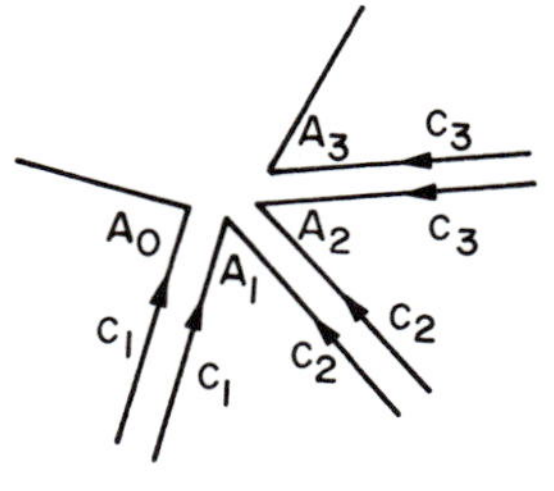

Fig. 3.18

Let T be a partial polyhedron.

After identification of the labeled sides let β_0 be the number of vertices, and β_1 the number of edges including the boundary edges. β_2 is the number of polygons of T. Then the *Euler characteristic* $E(T)$ of T is defined as $E(T)=\beta_0-\beta_1+\beta_2$.

Theorem 3.6. *If T is a partial polyhedron, then* $E(T)\leqq 1$.

Proof. The boundary pseudograph B of T has only vertices of valence 2. Therefore B consists of k closed ways and $k \geqq 1$. Let $s_1, s_2, \ldots, s_k$ be the lengths of these closed ways. Then take k additional polygons, namely, an s_1-gon, an s_2-gon, ..., and an s_k-gon. Identify the boundary of the s_i-gon with the corresponding closed way of boundary sides of T $(i=1, 2, \ldots, k)$.

The polyhedron P which is obtained has β_2+k polygons. By Theorem 3.2 it follows

$$E(P)=\beta_0-\beta_1+\beta_2+k=E(T)+k \leqq 2.$$

Since $k \geqq 1$ this completes the proof of Theorem 3.6.

The equation $E(T)=1$ holds if and only if $E(P)=2$ and $k=1$. Using Theorem 3.3 we obtain

Theorem 3.7. *If T is a partial polyhedron such that $E(T)=1$ then the boundary edges of T form one closed way and by identifying this way with the boundary of one additional polygon we obtain a subdivision of the sphere.*

Remark. One polygon with all sides unlabeled can also be considered as a partial polyhedron; we call it a 2-*cell.*

Each partial polyhedron T with the Euler characteristic $E(T)=1$ is elementarily related to a 2-cell. We do not need this theorem later. Therefore we omit the lengthy proof here.

4. Graphs on Surfaces

4.1. Embedding Theorem

Given a graph G and a surface S, one asks the following question. Is it possible to "draw". this graph G on the surface S such that the arcs of G intersect only at their common vertices? This concept must be precisely defined. Remember that a surface S is just a set of polyhedra. (See the definition in Section 3.6.)

Two graphs G, G' are called *isomorphic* if there exists a 1-1-correspondence between the set of vertices of G and the set of vertices of G' such that two vertices are adjacent in G if and only if the corresponding two vertices are adjacent in G'. For example the graph G_5 of Fig. 2.1 is isomorphic to the graph consisting of the vertices and edges of a triangular prism.

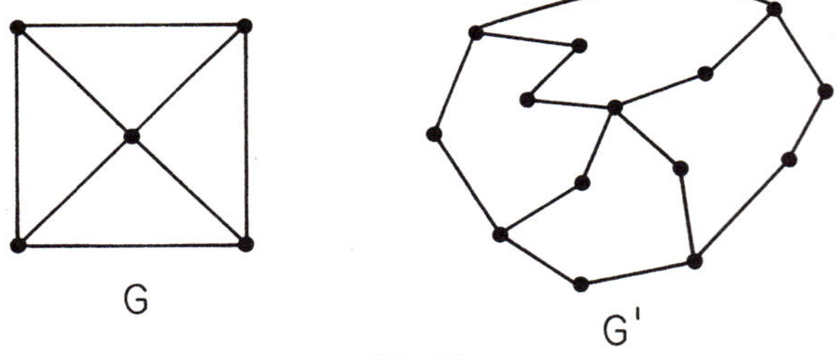

Fig. 4.1

Two graphs G and G' are said to be *homeomorphic* if G can be transformed into a graph isomorphic to G' solely by applying compositions and subdivisions of dimension one (as in Section 3.3) a finite number of times. Two homeomorphic graphs are illustrated in Fig. 4.1. In an abstract sense isomorphic graphs are identical whereas homeomorphic graphs have the same properties concerning embeddings.

We denote the numbers of vertices and arcs of G and G' in a natural way: $\alpha_0, \alpha_1, \alpha'_0, \alpha'_1$.

Theorem 4.1. *If the graphs G and G' are homeomorphic, then* $\alpha_0 - \alpha_1 = \alpha_0' - \alpha_1'$.

The proof is easy. In a subdivision of dimension one, one arc is replaced by two arcs and one additional vertex. Therefore the value of $\alpha_0 - \alpha_1$ is unchanged. The same is true for the reverse operation, a composition of dimension one.

Now we will define what an embedding ("drawing") of a graph G into ("on") a surface S is. If P is a polyhedron we define the pseudograph consisting of all the edges and vertices of P to be the 1-*skeleton* of P. For instance the 1-skeleton of the polyhedron $a b a^{-1} b^{-1}$ on the torus is a pseudograph with one vertex and two loops. Another example is the tetrahedron, whose 1-skeleton is K_4.

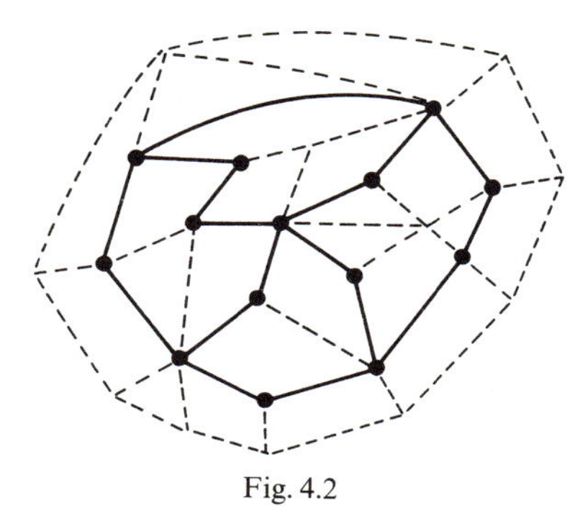

Fig. 4.2

Given a graph G and a surface S, we say that G can be embedded into S if there exists a polyhedron P on S such that the 1-skeleton of P has a subgraph homeomorphic to G (see Figs. 4.1 and 4.2). Intuitively this means that G can be drawn on the surface S without any arcs intersecting except at the vertices. A graph which can be embedded into the sphere is called a *planar graph*.

Theorem 4.2. *Let G be a graph where each vertex is of valence $\geqq 2$. If G can be embedded into a closed surface S then*

$$\alpha_1 \leqq 3\alpha_0 - 3E(S).$$

Since the proof of Theorem 4.2 is somewhat complicated, we postpone it and consider the following special case first.

Assume G happens to be the 1-skeleton of a polyhedron P on S. Then $E(S) = E(P) = \alpha_0 - \alpha_1 + \alpha_2$ where α_0 and α_1 are determined by G. We need an upper estimate for α_2. Since the shortest possible closed way in G has at least three arcs, each polygon of P has at least three edges. Moreover each edge of P is composed of two polygon sides.

Therefore $3\alpha_2 \leqq 2\alpha_1$ holds. This implies $3E(S) \leqq 3\alpha_0 - \alpha_1$ which is the inequality we wished to prove.

General Proof of Theorem 4.2. Assume G can be embedded into S. By definition there exists a polyhedron P on S and a subgraph G' of the 1-skeleton of P which is homeomorphic to G. Each edge of G' corresponds to two polygon sides in the plane representation of P. Therefore there are $2\alpha'_1$ polygon sides which correspond to the edges of G'. Now omit the labeling of these $2\alpha'_1$ polygon sides, i.e., no longer include the pairing (identification) for them. In doing this the polyhedron is broken down into partial polyhedra $T_1, T_2, \ldots, T_t$. (There may be more than one partial polyhedron if connectivity is disrupted.) We can describe this process as *cutting* the polyhedron along G'.

Let $\beta_0^{(i)}$ be the number of vertices, $\beta_1^{(i)}$ be the number of edges, and $\beta_2^{(i)}$ be the number of faces in T_i $(i=1, 2, \ldots, t)$. Let $\beta_0, \beta_1, \beta_2$ be these values for the polyhedron P.

Consider the three sums

$$\sum_{i=1}^{t} \beta_0^{(i)}, \quad \sum_{i=1}^{t} \beta_1^{(i)}, \quad \sum_{i=1}^{t} \beta_2^{(i)}.$$

In the first sum each vertex A of G' is counted exactly h times if $\operatorname{val} A = h$. For instance if $h=3$ the vertex A appears either once in three different partial polyhedra, in two polyhedra but twice in one, or in one partial polyhedron three times.

In the second sum each arc in G' is counted twice. It follows that

$$\beta_0 = \sum_{i=1}^{t} \beta_0^{(i)} - \sum_{A \in G'} \operatorname{val} A + \alpha'_0,$$

$$\beta_1 = \sum_{i=1}^{t} \beta_1^{(i)} - \alpha'_1,$$

$$\beta_2 = \sum_{i=1}^{t} \beta_2^{(i)}.$$

By Theorem 2.1 the sum of the valences of all vertices in the graph G' equals $2\alpha'_1$. We obtain:

$$\begin{aligned} E(S) &= \beta_0 - \beta_1 + \beta_2 \\ &= \sum_{i=1}^{t} (\beta_0^{(i)} - \beta_1^{(i)} + \beta_2^{(i)}) - 2\alpha'_1 + \alpha'_0 + \alpha'_1. \end{aligned}$$

Using Theorem 4.1 we obtain:

$$E(S) = \sum_{i=1}^{t} E(T_i) - \alpha_1 + \alpha_0. \tag{4.1}$$

Theorem 3.6 says that

(4.2) $$E(T_i) \leqq 1 \quad (i=1, 2, \ldots, t)$$

holds. Hence

(4.3) $$E(S) \leqq t - \alpha_1 + \alpha_0 .$$

Now let us compare t to α_1. Since G has no vertices of valence $\leqq 1$ each circuit induced by any rotation of G has at least length 3. Each arc in G is represented by a way in G'. (See Fig. 4.1.) Then the boundary pseudograph of each T_i represents at least three arcs of G. Each arc in G corresponds to two or one of the partial polyhedra T_i.

Therefore

(4.4) $$3t \leqq 2\alpha_1$$

and we obtain

$$3E(S) \leqq 3t - 2\alpha_1 - \alpha_1 + 3\alpha_0,$$

(4.5) $$3E(S) \leqq 3\alpha_0 - \alpha_1 .$$

This completes the proof of Theorem 4.2.

Now observe that equality holds in (4.5) only if equality holds in (4.4) *and* (4.2). This leads to the following

Theorem 4.3. *Let G be a graph, all of whose vertices have valence $\geqq 2$. If G can be embedded into the surface S and $\alpha_1 = 3\alpha_0 - 3E(S)$ holds, then there exists a polyhedron F on S such that G is isomorphic to the 1-skeleton of F and all faces of F are triangles.*

The proof is not easy. But since this theorem is of fundamental importance for this book we will give a complete proof:

Assume G can be embedded into the surface S. This means by definition that there is a polyhedron P on S and a subgraph G' of the 1-skeleton of P such that G' is homeomorphic to G. We will use all the notations and results given in the proof of the previous Theorem 4.2.

Because (4.5) is now assumed to be an equality, equality must hold in (4.2) and (4.4). Since for each i we have $E(T_i)=1$ we can apply Theorem 3.7. The boundary edges of T_i form one closed way. We identify this way with the boundary of one additional polygon L_i and obtain a subdivision D_i of the sphere.

It is necessary to "get rid" of all "superfluous" arcs of the 1-skeleton of P in order to construct a 1-skeleton that is isomorphic to G. For this reason apply the following process, first for $i=1$. Consider the plane representation of the polyhedron P and replace all the polygons of T_1 by the single polygon L_1 such that the boundary circle of L_1 corresponds to the boundary of T_1.

This will produce a new polyhedron P' on a surface S'. We will show that $S' = S$. In order to determine the Euler characteristic of S' we use the already proven formula (4.1). This time the first partial polyhedron is not T_1 but L_1. However, $E(L_1) = E(T_1) = 1$ and therefore $E(S') = E(S)$.

Assume P' is orientable. Then for each polygon of P' one can choose an orientation such that each edge of P' is used in both directions. We call such a choice an *orientation* of P'. Each orientation (there are two different ones) of P' is determined by the orientation of one of the polygons.

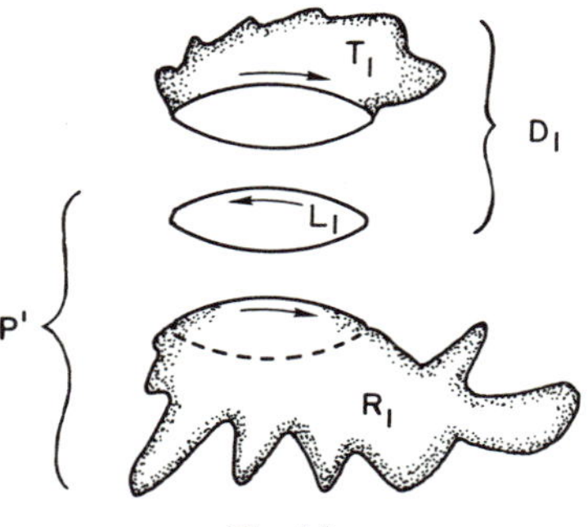

Fig. 4.3

Denote the partial polyhedron consisting of the union of $T_2, T_3, \ldots, T_t$ by R_1. Choose an orientation for P'. This determines an orientation for L_1 and R_1. Then choose an orientation for D_1 determined by the given orientation of L_1. This gives an orientation for T_1. The opposite orientation of T_1 together with the given orientation of R_1 gives an orientation for P. (See Fig. 4.3.) In the same way one can see that P' is orientable if P is. Thus P and P' are both orientable or both non-orientable. Therefore, by Theorem 3.4, the two surfaces S and S' are identical.

Now consider the polyhedron P and replace $T_1, T_2, \ldots, T_t$ consecutively by the polygons $L_1, L_2, \ldots, L_t$. We get a polyhedron on S which has a 1-skeleton isomorphic to G'. After some subdivisions or compositions of dimension one we obtain a polyhedron F on S which has a 1-skeleton isomorphic to G. Since equality holds in (4.4) all the faces in F are triangles. This completes the proof of Theorem 4.3.

If a graph G satisfies the conditions of Theorem 4.3 the polyhedron F on S is called a *triangular embedding* of G into S. For instance K_3 and K_4 have triangular embeddings into the sphere and K_7 is triangularly embeddable into the torus (Fig. 1.7).

A triangular embedding of a graph $\neq K_3$ has the property that any two different triangles have either no elements, just one vertex, or just one arc and two vertices in common. Otherwise there would be loops or parallel arcs in the graph.

The following theorem is the converse of Theorem 4.3.

Theorem 4.4. *If there exists a triangular embedding of a graph G into a surface S, then* $\alpha_1 = 3\alpha_0 - 3E(S)$.

Denote the number of triangles in this embedding by α_2. Then $3\alpha_2 = 2\alpha_1$. Substituting this into Euler's formula $E(S) = \alpha_0 - \alpha_1 + \alpha_2$, we obtain Theorem 4.4.

The following generalization of Theorem 4.2 is sometimes very useful.

Theorem 4.5. *Let G be a graph with vertices all of valence* $\geqq 2$ *and let the shortest closed way in G be of length* $\geqq k$. *If G is embeddable into the closed surface S then*

$$(k-2)\,\alpha_1 \leqq k\,\alpha_0 - k\,E(S).$$

The proof is a slight modification of the proof for Theorem 4.2: Instead of (4.4), consider the inequality

$$k\,t \leqq 2\alpha_1$$

and substitute this into inequality (4.3) to obtain the result.

Exercises

4.1.1. Show that the complete graph K_5 is not planar.

4.1.2. Do the same for the graphs G_6 and G_7 in Fig. 2.1.

4.1.3. The following graph is called the *complete bipartite graph* $K_{n,m}$ with $n+m$ vertices. There are two disjoint classes with n and m vertices respectively, and any two vertices in different classes are joined by one arc. So $K_{n,m}$ has $n\,m$ arcs. For instance the graph G_6 in Fig. 2.1 is isomorphic to $K_{3,3}$. Show that $K_{4,4}$ is not embeddable into the non-orientable surface of genus one.

4.1.4. Assume the graph $K_{n,m}$ defined in Exercise 4.1.3. is embeddable in the surface S. Find an upper bound for $E(S)$.

4.1.5. Theorem 4.2 holds not only for graphs with vertices of valence $\geqq 2$. It is enough to assume that the number of vertices is at least 3. Prove this.

4.1.6. Somebody has drawn a pseudograph on the plane with 11 vertices and 28 arcs without intersecting arcs except at the vertices. Show that this is not a graph.

4.1.7. Given a surface S, show that there exists a graph which cannot be embedded into S.

4.1.8. A graph is called *hexavalent* if each vertex has valence 6. Assume that a hexavalent graph can be embedded into the surface S. Prove that S is not the sphere and not the non-orientable surface of genus 1.

4.2. Dual Polyhedra

A polyhedron M on a surface is called a *map on S* if each vertex of M has valence $\geqq 3$ and each vertex of valence h is incident with h different faces. The faces of a map are called *countries*. Two countries with at least one common edge are called *adjacent* countries. The definition of a map prevents a country from being adjacent to itself (along an edge). But it can happen that two countries have more than one edge in common. The chromatic number $\chi(M)$ of a map M has already been defined in Chapter 1. Since we know much more about graphs than about maps, we shall concern ourselves with graphs associated to maps.

For each map M we define the following abstract graph G_M and call it the *country graph* of M. Each country of M is represented by one vertex of G_M and two vertices are adjacent in G_M if and only if the two corresponding countries are adjacent in M. It is obvious that $\chi(G_M)=\chi(M)$, for each coloring of the countries of M can immediately be translated into a coloring of the vertices in G_M and vice versa.

Theorem 4.6. *If M is a map on the surface S then the country graph G_M can be embedded into S.*

There is a standard proof of this theorem: Choose a point in the interior of each country and call it the *capital* of the country. Then join any two capitals by a blue arc (some constructions can be more easily explained if one uses "colored arcs" which only means "certain arcs"), if the corresponding countries are adjacent (see Fig. 4.4). All blue arcs can be chosen on the surface such that they do not intersect each other except at the capitals. A subgraph of the pseudograph of all capitals and all blue arcs is isomorphic to the graph G_M. (A pair of countries can have more than one common edge but we need only one arc joining the two capitals.) This proof is questionable because it is addressed only to intuition and does not fit into the theory of surfaces. But it will become precise if we copy the above construction by using *only* elementary operations (subdivisions of dimension one and two).

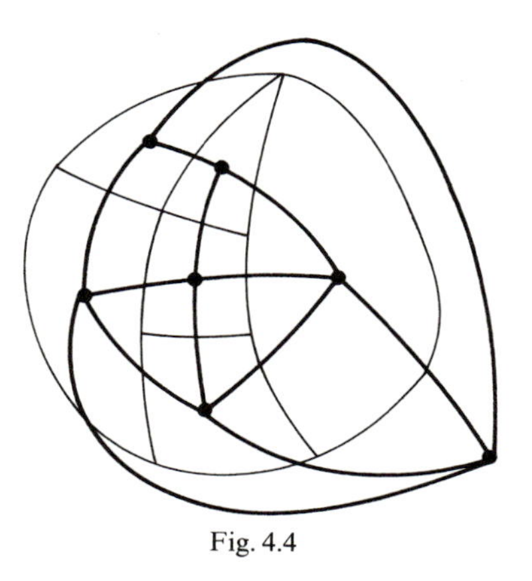

Fig. 4.4

Given a map M on a surface S, first divide each edge of M into two edges by adding a new vertex (subdivision of dimension 1). Call these new vertices *midpoints*. Since each country is adjacent to at least two other countries, each country is incident with at least two midpoints. Consider a country λ and join two of the midpoints incident with λ by a new arc entirely contained in λ. Thus λ is divided into two pieces (subdivision of dimension 2). Divide this new arc into two arcs by a new vertex which we will call the *capital* of λ (subdivision of dimension 1). Then join each of the other midpoints, originally incident with λ, one after the other with the capital of λ by a new arc (subdivision of dimension 2). Proceed in such a manner for each country.

For the polyhedron obtained on S, the capitals and all the new arcs form a pseudograph which contains a subgraph homeomorphic to G_M. This completes the proof of Theorem 4.6.

If each pair of adjacent countries in the map M has only one edge in common, this construction can be continued. Applying compositions of dimension 2 and 1 we can get rid of all original edges and vertices of M. We can then delete the midpoints as well. The 1-skeleton of the map obtained on S is isomorphic to the graph G_M. This map is called the *dual map* of M.

There are some well known examples of pairs of dual maps on the sphere:

the cube and the octahedron;
the dodecahedron and the icosahedron;
the tetrahedron is dual to itself.

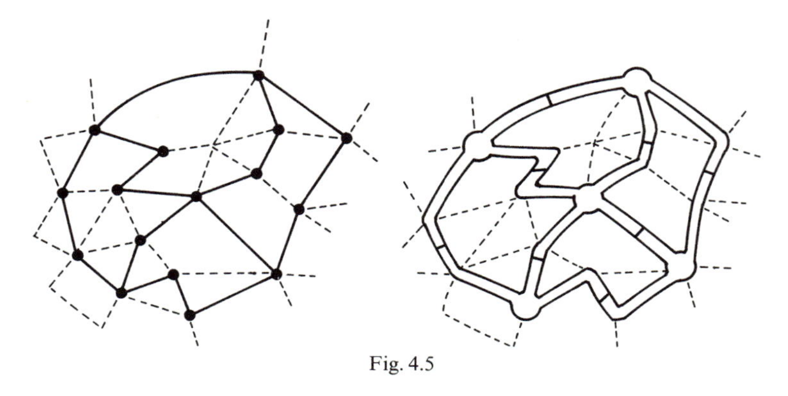

Fig. 4.5

On the torus, the map in Fig. 1.3 is dual to the map in Fig. 1.7 and vice versa. (There is also a selfdual map on the torus consisting of five quadrilaterals.)

Theorem 4.7. *If the graph G can be embedded into the surface S then there exists a map M on S such that G is isomorphic to a subgraph of the country graph G_M.*

The following proof can easily be completed rigorously by using only subdivisions and compositions of dimensions 1 and 2 (as in the proof for Theorem 4.6). But we will only outline the procedure.

Given an embedding of G into S, construct a "little circle" around each vertex of G (Fig. 4.5). Expand each arc of G (it might be represented

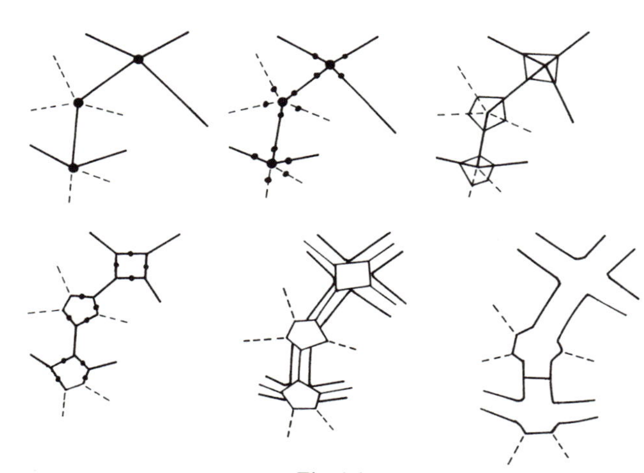

Fig. 4.6

by a way on S) to a "narrow street". Using a new arc, divide each of these streets into two halves. Consider each circle including all of its adjoined "halfstreets" as a country. Let the rest of the surface be divided into some more countries.

The country graph of the obtained map M certainly contains a subgraph isomorphic to the given graph G. Fig. 4.6 illustrates how to do this construction by using only elementary operations.

4.3. Heawood's Inequality

If M is a map on a surface S we sometimes use the notation $M \subset S$. Similarly, the notation $G \subset S$ means, by definition, that the graph G can be embedded into S.

In Chapter 1, the chromatic number $\chi(S)$ of a closed surface S is defined as

$$\chi(S) = \max_{M \subset S} \chi(M) \tag{4.6}$$

where the maximum is taken over all possible maps M on S. At the moment it is not clear whether this maximum in (4.6) exists or not. But the following Theorem 4.8 shows that the maximum in (4.6) and also in (4.7) is a finite number. It is now easy to show that we can take graphs instead of maps. We now prove that

$$\chi(S) = \max_{G \subset S} \chi(G) \tag{4.7}$$

where the maximum is taken over all graphs G which can be embedded into S.

For each map M on S there exists a graph, namely, the country graph G_M, embeddable (Theorem 4.6) into S such that $\chi(M) = \chi(G_M)$. Thus

$$\max_{M \subset S} \chi(M) \leqq \max_{G \subset S} \chi(G).$$

The reverse inequality follows immediately from the fact that for each graph G on S there exists a map M on S (Theorem 4.7) with $\chi(G) \leqq \chi(M)$.

We are now going to determine the chromatic number of a closed surface. The first step is the following inequality which was found in 1890 by Heawood. His proof was not too precise and he considered only orientable surfaces.

Theorem 4.8. *If S is a closed surface with Euler characteristic $E(S) \neq 2$ (this means S is different from the sphere) then*

$$\chi(S) \leqq \left[\frac{7 + \sqrt{49 - 24E(S)}}{2}\right]. \tag{4.8}$$

Proof. By virtue of (4.7) there exists a graph G which can be embedded into S such that $\chi(S)=\chi(G)$. We can assume that G is critical (otherwise consider a critical subgraph with the same chromatic number). By Theorems 2.4 and 4.2 it follows for $\chi(G)=\chi$ and $E(S)=E$ that

$$(\chi-1)\,\alpha_0 \leqq 2\alpha_1, \qquad \alpha_1 \leqq 3\alpha_0-3E,$$

$$(\chi-1)\,\alpha_0 \leqq 6\alpha_0-6E,$$

$$\chi-1 \leqq 6-\frac{6E}{\alpha_0}. \tag{4.9}$$

We now consider the two cases: $E\leqq 0$ or $E=1$.

Case I. Assume $E\leqq 0$. Since $\alpha_0 \geqq \chi$ we can replace α_0 by χ in (4.9) and multiply by χ giving

$$\chi^2-\chi \leqq 6\chi-6E,$$

$$\chi^2-7\chi+6E \leqq 0.$$

Using the quadratic formula this can be written in the form

$$\left(\chi-\frac{7+\sqrt{49-24E}}{2}\right)\left(\chi-\frac{7-\sqrt{49-24E}}{2}\right) \leqq 0. \tag{4.10}$$

Since $E\leqq 0$ it follows that $\sqrt{49-24E}\geqq 7$. Since $\chi\geqq 1$, the second factor of (4.10) is always positive. It follows that the first factor is $\leqq 0$. This completes the proof of Theorem 4.8 in case $E\leqq 0$.

Case II. Assume $E=1$, i.e., S is the non-orientable surface of genus 1. Then from (4.9) we obtain

$$\chi-1 \leqq 6-\frac{6}{\alpha_0} < 6.$$

Since χ is an integer it follows that

$$\chi \leqq 6 = \frac{7+5}{2} = \frac{7+\sqrt{49-24E}}{2}.$$

There is really a third case, $E=2$, to consider, but a proof or a counterexample for this case is unknown. This case is of course the celebrated four color problem.

The next target will be to prove that the inequality of Theorem 4.8 really holds as an equality (with one exception). For this purpose we have to exhibit a graph G on a given surface S such that $\chi(G)$ equals the right side of (4.8).

Fortunately the complete graph is appropriate for doing the job. For other graphs, it would be rather difficult to determine their chromatic number. We therefore have to study the following question: *Given a surface S, for what n does there exist an embedding of K_n into S?* For practical purposes it is better to ask the same question in the following form: *Given a number n, which surfaces S have the property that K_n can be embedded into S?*

We will study this question separately for orientable and for non-orientable surfaces.

4.4. Genus of Graphs

The *genus* $\gamma(G)$ of a graph G is defined as the minimal genus of an orientable surface into which G can be embedded. In other words $\gamma(G)=p$ means that G can be embedded into S_p but not into S_{p-1}. For instance $\gamma(K_4)=0$ and $\gamma(K_5)=1$. Assume G has no vertices of valence $\leqq 1$. Using Theorem 4.2 we obtain

$$\alpha_1 \leqq 3\alpha_0 - 6 + 6p,$$

$$\alpha_1 - 3\alpha_0 + 6 \leqq 6p,$$

$$\gamma(G) \geqq \left\{\frac{\alpha_1 - 3\alpha_0 + 6}{6}\right\}. \tag{4.11}$$

The symbol $\{x\}$ means the smallest integer $\geqq x$.

The Thread Problem in Section 1.3 really means the determination of the genus of the complete graph. The solution has two parts, one is very easy, the other very difficult. Consider first the easy part.

Applying (4.11) with $G=K_n$, $\alpha_0=n$, $\alpha_1=\binom{n}{2}$ we get:

$$\gamma(K_n) \geqq \left\{\frac{(n-3)(n-4)}{12}\right\} \quad \text{for } n \geqq 3. \tag{4.12}$$

In some of the next chapters we will give a proof for the fact that (4.12) *really holds as an equality:*

$$\gamma(K_n) = \left\{\frac{(n-3)(n-4)}{12}\right\} \quad \text{for } n \geqq 3. \tag{4.13}$$

The proof is divided into 12 cases depending on the residue class of n (mod 12).

For $n \equiv 0$, 3, 4 and 7 (mod 12) the product $(n-3)(n-4)$ is divisible by 12. These four cases are called the *regular* cases because for these numbers n one is able to construct a triangular embedding of K_n into an orientable surface of genus $\gamma(K_n)$. We will do this later.

If the genus of K_n can be determined by proving (4.13) the Map Color Theorem can easily be shown for all orientable surfaces except the sphere.

Theorem 4.9. *If the Eq.* (4.13) *is true, then*

$$\chi(S_p) = \left[\frac{7+\sqrt{1+48p}}{2}\right] \quad \text{for each } p \geqq 1.$$

Proof. Given the orientable surface S_p of genus $p \geqq 1$, let n be the largest integer such that

$$\gamma(K_n) = \left\{\frac{(n-3)(n-4)}{12}\right\} \leqq p. \tag{4.14}$$

Then

$$p < \left\{\frac{(n-2)(n-3)}{12}\right\},$$

$$12p < (n-2)(n-3),$$

$$0 < n^2 - 5n + 6 - 12p,$$

$$0 < \left(n - \frac{5+\sqrt{1+48p}}{2}\right)\left(n - \frac{5-\sqrt{1+48p}}{2}\right).$$

Since $p \geqq 1$ the second factor is always positive. Therefore

$$0 < n - \frac{5+\sqrt{1+48p}}{2},$$

$$\frac{7+\sqrt{1+48p}}{2} - 1 < n. \tag{4.15}$$

From (4.14) we have $\gamma(K_n) \leqq p$. Therefore K_n can be embedded into S_p which means that $n \leqq \chi(S_p)$. From (4.15) we obtain

$$\frac{7+\sqrt{1+48p}}{2} - 1 < \chi(S_p) \leqq \frac{7+\sqrt{1+48p}}{2}. \tag{4.16}$$

The upper bound on the right hand side of (4.16) is obtained from Theorem 4.8 with $E(S_p) = 2 - 2p$. Theorem 4.9 now follows immediately from (4.16).

4.5. Non-Orientable Genus of Graphs

Let us consider the same questions for non-orientable surfaces. The *non-orientable genus* $\bar{\gamma}(G)$ of a graph G is defined as the minimal genus of a non-orientable surface into which G can be embedded. Assume that all the vertices of G are of valence $\geqq 1$. Using Theorem 4.3 we obtain

$$\alpha_1 \leqq 3\alpha_0 - 3E(S) \quad \text{with} \quad E(S) = 2 - \bar{\gamma}(G),$$

$$\bar{\gamma}(G) \geqq \left\{ \frac{\alpha_1 - 3\alpha_0 + 6}{3} \right\}. \tag{4.17}$$

For the complete graph, this yields the formula

$$\bar{\gamma}(K_n) \geqq \left\{ \frac{(n-3)(n-4)}{6} \right\} \quad \text{for } n \geqq 3. \tag{4.18}$$

In some of the later chapters we will establish a proof of the following equation

$$\bar{\gamma}(K_n) = \left\{ \frac{(n-3)(n-4)}{6} \right\} \quad \text{for } n \neq 7 \text{ and } n \geqq 3, \tag{4.19}$$

$$\bar{\gamma}(K_7) = 3.$$

The case $n=7$ is a remarkable exception. (See Section 4.6, Klein's bottle.)

Theorem 4.10. *If Eq.* (4.19) *is true then*

$$\chi(N_q) = \left[\frac{7 + \sqrt{1 + 24q}}{2} \right] \quad \text{for } q \neq 2.$$

The exceptional case N_2 (Klein's bottle) will be handled separately. The proof of Theorem 4.10 is almost the same as that of Theorem 4.9.

However, we must first consider the special case $q=3$. In this case we can prove Theorem 4.10 even without using Eq. (4.19). From (4.8) we obtain

$$\chi(N_3) \leqq \left[\frac{7 + \sqrt{73}}{2} \right] = 7.$$

Fig. 1.12 illustrates a map with chromatic number 7 on M_3 which can be considered as a part of the surface N_3. Therefore $7 \leqq \chi(N_3)$ and Theorem 4.10 is proved for $q=3$.

For $q=1$ we can use the same argument to prove $\chi(N_1)=6$ on account of (4.8) and Fig. 1.9. (But we would *not* be successful in case $q=2$ just using (4.8) and Fig. 1.11.)

Now we are ready for the general proof of Theorem 4.10. Assume (4.19) is true and let q be an integer $\geqq 4$. Choose the largest integer n such that

$$\left\{ \frac{(n-3)(n-4)}{6} \right\} \leqq q.$$

Then $n \geqq 8$ and applying (4.19) we obtain

$$\bar{\gamma}(K_n) \leqq q. \tag{4.20}$$

Because n is maximal we get

$$q < \frac{(n-2)(n-3)}{6},$$

$$6q < n^2 - 5n + 6,$$

$$0 < n^2 - 5n + 6 - 6q,$$

$$0 < \left(n - \frac{5 + \sqrt{1+24q}}{2}\right)\left(n - \frac{5 - \sqrt{1+24q}}{2}\right).$$

The second factor is positive and therefore

$$0 < n - \frac{5 + \sqrt{1+24q}}{2}$$

$$\frac{7 + \sqrt{1+24q}}{2} - 1 < n.$$

By (4.20), $\bar{\gamma}(K_n) \leqq q$ and therefore K_n is embeddable into the surface N_q. This means $n \leqq \chi(N_q)$. From the formula above it follows that

$$\frac{7 + \sqrt{1+24q}}{2} - 1 < \chi(N_q) \leqq \frac{7 + \sqrt{1+24q}}{2}. \tag{4.21}$$

The last part of inequality (4.21) is obtained from Theorem 4.8 with $E(N_q) = 2 - q$. The equality of Theorem 4.10 now follows from (4.21).

Exercises

4.5.1. Prove that the chromatic number of the sphere is less than or equal to six.

4.5.2. Find a graph for which (4.11) does not hold as an equality.

4.5.3. Show that for each G the inequality $\bar{\gamma}(G) \leqq 2\gamma(G) + 1$ is true.

4.5.4. Find the genus of the graph $K_n - K_{n-2}$.

4.5.5. Consider a regular $2n$-gon with all the $2n$ sides and all the largest diagonals (their number is n) as a graph. What is the genus of this trivalent graph? (It is smaller than normally expected.)

4.5.6. What is the non-orientable genus of the graph in Exercise 4.5.5?

4.5.7. Consider a regular $(2n+1)$-gon with all the $2n+1$ sides and all the largest diagonals (there are $2n+1$ large diagonals) as a graph. What is the genus of this tetravalent graph?

4.5.8. Consider a regular n-gon with all the n sides and all the smallest diagonals (longer than a side) as a graph. What is the genus of this tetravalent graph?

4.5.9. What is the genus of the bipartite graph $K_{4,4}$? (For definition see 4.1.3.)

4.5.10. Determine the genus of $K_7 - K_4$.

4.6. Klein's Bottle

Now we will study the surface N_2 which is also called Klein's bottle. This special case is not covered by Theorem 4.10 and is particularly instructive.

Theorem 4.11. *The complete graph K_7 cannot be embedded into the surface N_2.*

Proof. Assume K_7 can be embedded into N_2. Since $\alpha_0=7$, $\alpha_1=21$, and $E(N_2)=0$ the equation $\alpha_1=3\alpha_0-3E(N_2)$ holds. Therefore Theorem 4.3 is applicable, i.e., there exists a triangular embedding of K_7 into N_2. The dual map M of this polyhedron consists of seven mutually adjacent hexagons and all vertices have valence 3.

We are now concerned with the structure of the map M. Consider a vertex P. Label the three hexagons incident with the vertex P by 0, 1, 2. The edge between hexagon 1 and hexagon 2 touches another vertex $Q \neq P$. This vertex Q is incident with hexagon 1, hexagon 2, and a third one which may be called 3. Hexagon 3 is different from hexagon 0, otherwise hexagon 1 would not have enough edges to be adjacent to 6 different hexagons. In the same way we see that hexagons 0, 1, 2, 3, 4, 5, as in Fig. 4.7, are all different from each other. The last hexagon labeled

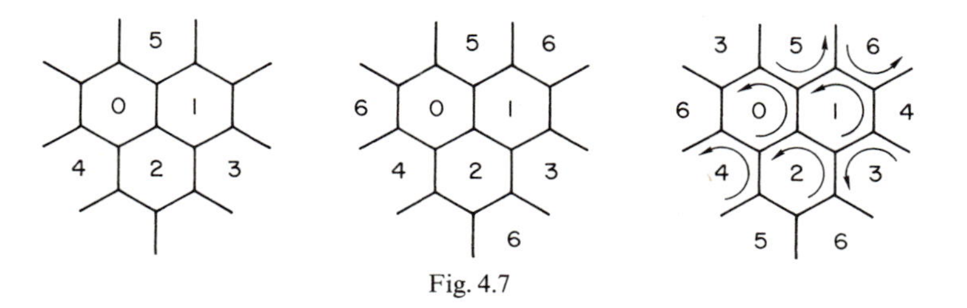

Fig. 4.7

6 must be located as in the central part of Fig. 4.7 or else hexagon 6 coincides with the three unlabeled places. The two possibilities are not really different from each other. One is the mirror image of the other (reflection on the central vertical line).

Now 0 must be adjacent to 3, and 1 to 4, and 2 to 5. Therefore a partial picture of the map M *must* look like the right hand side of Fig. 4.7. We will soon see that this partial picture uniquely determines the entire map M.

For each hexagon, arbitrarily choose an orientation. Then write down the cyclic order of the six neighbors of hexagon 0 given by the chosen orientation: 4 2 1 5 3 6. Do the same for the other hexagons. For hexagon 3 you get the partial orders 4 1 2 6 and 6 0 5 or 5 0 6. But since 6 appears only once the total order has to read 4 1 2 6 0 5.

The partial picture of Fig. 4.7 determines the complete list of those cyclic orders:

(4.22)

0.	4	2	1	5	3	6
1.	2	3	4	6	5	0
2.	3	1	0	4	5	6
3.	4	1	2	6	0	5
4.	5	2	0	6	1	3
5.	3	0	1	6	2	4
6.	5	1	4	0	3	2

From Fig. 4.8 we see that in the scheme (4.22) the following rule must be satisfied.

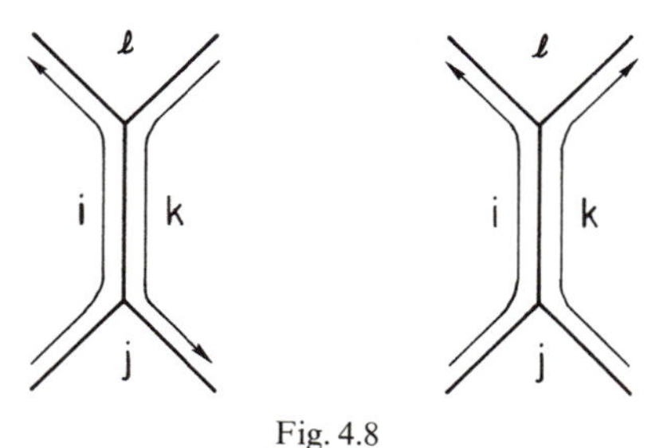

Fig. 4.8

Rule *R*. *If* $i. \ldots jkl \ldots$ *appears in line* i, *then* $k. \ldots lij \ldots$ *or* $k. \ldots jil \ldots$ *must appear in line* k.

One can check that rule R is true for each pair i, k in (4.22). So the assumption is still uncontradicted. But if we check once more we can easily see that the even stronger rule R^* is true (see also Section 2.3).

Rule R^*. *If* $i.\ \ldots j\,k\,l \ldots$ *then* $k.\ \ldots l\,i\,j \ldots$.

But this means that the map M is orientable which is a contradiction. So Theorem 4.11 is proved. The scheme (4.22) is really the same as (2.7) after an appropriate renumbering.

We have seen that K_7 cannot be embedded in N_2. To see that K_7 can be embedded in N_3, look at Fig. 1.12. Since all seven countries of the map are mutually adjacent, the 1-skeleton of the dual map contains K_7 as a subgraph. But Fig. 1.12 is a map on M_3, which is a part of N_3. So K_7 can be embedded in N_3. This means $\bar{\gamma}(K_7)=3$, which proves (4.19) if $n=7$. Alternatively, since K_7 can be embedded in the torus S_1, by adding one crosscap to the surface we obtain an embedding of K_7 in N_3.

Before determining the chromatic number of Klein's bottle we need to explain a graph theoretical operation. Let G be a graph and c an arc of G with incident vertices A and B. Then consider the following graph G'. The vertex set of G' consists of the vertices of $G-A-B$ and one new vertex C. The arcs of G' are the arcs of $G-A-B$ and the new arcs PC for each P with the property that P and A or P and B are adjacent in G. (See Fig. 4.9.)

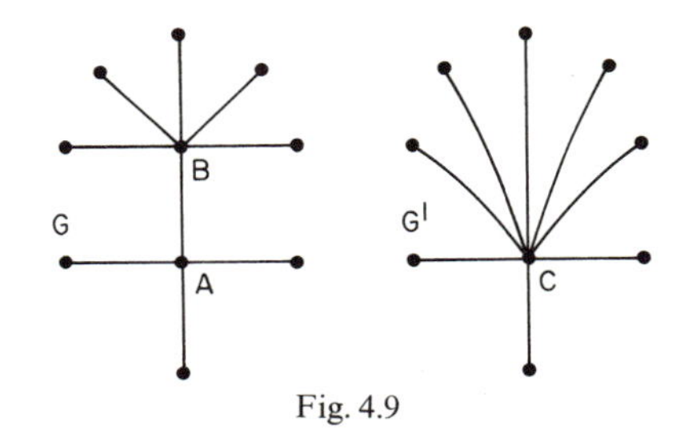

Fig. 4.9

We say that G' is obtained from G by *contraction* of the arc c. The number of arcs in G' is less than the number of arcs in G. (If P is adjacent to A *and* B in G there is only *one* arc from P to C in G'.)

Assume that the graph G can be embedded into a surface S. The contraction can easily be constructed on the surface so that G' can also be embedded into S. One can do this using elementary operations similar to those in Fig. 4.6.

Theorem 4.12. *The chromatic number of Klein's bottle is 6.*

Proof. If we apply Theorem 4.8 with $E(N_2)=0$ we only get the information that

$$\chi(N_2) \leqq 7.$$

Since the complete graph K_6 can be embedded into N_1 (Fig. 1.9), the graph K_6 can also be embedded into N_2. Therefore

$$6 \leqq \chi(N_2).$$

Now assume that $\chi(N_2)=7$. This means that there exists a graph embedded in N_2 with chromatic number 7. We can assume the graph has a minimal number of vertices. Now we consider a critical subgraph G, which means we still have $\chi(G)=7$. Denote the number of vertices and arcs in G by α_0 and α_1. Since G is embedded in N_2, Theorem 4.2 shows that

$$\alpha_1 \leqq 3\alpha_0.$$

Since G is critical with chromatic number $\chi=7$, Theorem 2.4 shows that $(\chi-1)\,\alpha_0 \leqq 2\alpha_1$ which means

$$3\alpha_0 \leqq \alpha_1.$$

Therefore

$$3\alpha_0 = \alpha_1. \tag{4.23}$$

By Theorem 2.3 the valence of each vertex of G is at least 6. Therefore, using Theorem 2.1

$$6\alpha_0 = \sum_{i=1}^{\alpha_0} 6 \leqq \sum_{i=1}^{\alpha_0} \operatorname{val} P_i = 2\alpha_1,$$

and equality can hold only if $\operatorname{val} P_i = 6$ for all i. Since (4.23) says that equality does hold, each vertex of G has exactly valence 6. We say G is *hexavalent*.

Consider a vertex P of G and its neighbors $Q_1, Q_2, Q_3, Q_4, Q_5, Q_6$. There must be a pair, say Q_1, Q_3, which is not adjacent in G. Otherwise we would have K_7 on the surface N_2 in contradiction to Theorem 4.11. Now contract first the arc PQ_1 and then the arc PQ_3 (see Fig. 4.10). The resulting graph G' can still be embedded into N_2 and has less vertices than G. Therefore $\chi(G') \leqq 6$ since G had minimal α_0.

Consider a coloration of the vertices of G' with 6 colors. Now it is easy to find a coloration of the original graph G. Just give Q_1 and Q_3

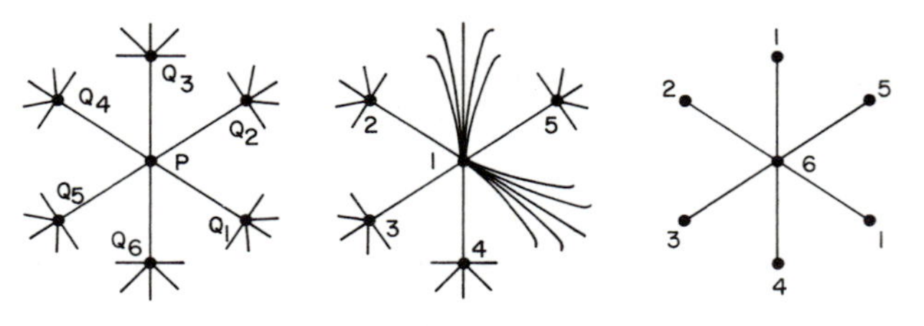

Fig. 4.10

the same color which P had in G'. The other vertices different from P get exactly the same color they had in G'. Then the 6 neighbors of P are colored with 5 or less different colors (Q_1 and Q_3 have the same color). So the sixth color can be assigned to P. We have found a coloration of G with six colors. *Therefore the chromatic number of* N_2 *is* 6.

Exercises

4.6.1. Prove that each graph G which can be embedded into the torus with $\chi(G)=7$ contains a subgraph isomorphic to K_7.

4.6.2. Let G be a graph and G' a graph obtained from G by a contraction of an arc. If G can be embedded into a surface S then G' can also be embedded into S. Show that the converse statement is false by giving a counterexample.

4.6.3. If a graph has only vertices of valence 5 or 6 and the number of vertices of valence 5 is smaller than 12, then the graph is not planar. Prove it.

5. Combinatorics of Embeddings

5.1. Triangular Embeddings

In order to prove the Map Color Theorem we only have to determine the genus and the non-orientable genus of the complete graph K_n, according to Theorems 4.9 and 4.10. That means we just have to prove Eqs. (4.13) and (4.19). Denote the right hand side of (4.13) by p and that of (4.19) by q. Then we have to exhibit an embedding of K_n into S_p and into N_q. The following table shows the values of p and q respectively for small values of n.

n	7	8	9	10	11	12	13	14	15	16	17	18	19	20	21	22
p	1*	2	3	4	5	6*	8	10	11*	13*	16	18	20*	23	26	29

n	6	8	9	10	11	12	13	14	15	16	17	18	19	20	21	22
q	1*	4	5*	7*	10	12*	15*	19	22*	26*	31	35*	40*	46	51*	57*

In order to get lower bounds for $\gamma(K_n)$ and $\bar{\gamma}(K_n)$ we have to construct embeddings such as $K_8 \subset S_2$, $K_9 \subset S_3$ and so on and $K_8 \subset N_4$, $K_9 \subset N_5$ and so on. As we did for $K_7 \subset S_1$ we could use a picture to represent the four mentioned embeddings. But because n is an arbitrary integer and can have high values we certainly need another method to prove the existence of such embeddings.

A star in the above table means that the corresponding number n has the property that $(n-3)(n-4)$ is divisible by 12 in the orientable case and that $(n-3)(n-4)$ is divisible by 6 in the non-orientable case. This happens only if

(5.1) $$n \equiv 0, 3, 4, \text{ or } 7 \pmod{12}$$

for orientable surfaces and if

(5.2) $$n \equiv 0 \text{ or } 1 \pmod 3$$

for non-orientable surfaces. These are the *regular* cases.

By the notation $G \lhd S$ we mean that there exists a triangular embedding of the graph G into the surface S.

For the regular cases equation $\alpha_1 = 3\alpha_0 - 3E(S)$ holds. Therefore Theorem 4.3 says:

If $K_n \subset S_p$ with $p = (n-3)(n-4)/12$ then $K_n \lhd S_p$.

If $K_n \subset S_q$ with $q = (n-3)(n-4)/6$ then $K_n \lhd N_q$. That means in the regular case we have to find a triangular embedding of K_n into an orientable surface S_p and into a non-orientable surface N_q. Then Theorem 4.4 implies that $p = (n-3)(n-4)/12$ and $q = (n-3)(n-4)/6$. In all the other cases we are sure that a triangular embedding does not exist. As we can see (by 5.1 and 5.2) there are "more" regular cases for non-orientable than for orientable surfaces. This makes the non-orientable surfaces much simpler to manage.

Now we explain a simple but powerful method to describe a given triangular embedding of a graph combinatorially.

Assume $K_n \lhd S$. Let S be an arbitrary orientable or non-orientable surface. Then the dual map of this embedding consists of n countries. Each is adjacent to every other along exactly one edge, therefore each country is an $(n-1)$-gon. Moreover each vertex is of valence three. Denote the countries by $0, 1, \ldots, n-1$. Then choose a certain orientation for each country. Write down the cyclic order of the countries adjacent to country 0. (We did the same in Section 4.6.) This gives a certain permutation of $1, 2, \ldots, n-1$. Do the same for the other countries. This leads to a scheme. The scheme (5.3) gives an example with $n=6$.

$$
\begin{array}{llllll}
0. & 3 & 1 & 5 & 4 & 2 \\
1. & 4 & 2 & 5 & 0 & 3 \\
2. & 0 & 3 & 5 & 1 & 4 \\
3. & 1 & 4 & 5 & 2 & 0 \\
4. & 2 & 0 & 5 & 3 & 1 \\
5. & 0 & 1 & 2 & 3 & 4
\end{array}
\tag{5.3}
$$

The scheme obtained can be considered as a certain rotation of K_6 or in general of K_n. Recall that we have described rotations of graphs in Section 2.2 by schemes. Assume that we have obtained the scheme (5.3) from an embedding $K_6 \lhd S$ in the described way. Fig. 4.8 shows that in the scheme (5.3) the rule R of Section 4.6 holds. Of course this is true not only in the special case $n=6$.

So we have shown

Theorem 5.1. *If $K_n \lhd S$ then there exists a rotation of K_n such that in the scheme describing this rotation Rule R holds.*

Much more important is the following converse of this theorem.

Theorem 5.2. *If in the scheme describing a rotation of K_n the Rule R holds then there exists a suitable surface S such that $K_n \lhd S$.*

Proof. Assume there is a rotation of K_n satisfying Rule R. We explain the proof for the example $n=6$. Given the scheme (5.3), we then write the following table

$$\begin{array}{ccccc} c_{03} & c_{01} & c_{05} & c_{04} & c_{02} \\ c_{14} & c_{12} & c_{15} & c_{10} & c_{13} \\ c_{20} & c_{23} & c_{25} & c_{21} & c_{24} \\ c_{31} & c_{34} & c_{35} & c_{32} & c_{30} \\ c_{42} & c_{40} & c_{45} & c_{43} & c_{41} \\ c_{50} & c_{51} & c_{52} & c_{53} & c_{54} \end{array} \tag{5.4}$$

In (5.4) the first index of the letter c is the number of the row. The second index is the number which appears in the corresponding place in the scheme in the row i in (5.3) ($i=0, 1, \ldots$).

We define

$$c_{ik}=c_{ki}, \quad \text{if in (5.3) there is } \begin{cases} i. \ \ldots j\,k\,l \ldots \\ k. \ \ldots j\,i\,l \ldots \end{cases}$$

$$c_{ik}=c_{ki}^{-1}, \quad \text{if in (5.3) there is } \begin{cases} i. \ \ldots j\,k\,l \ldots \\ k. \ \ldots l\,i\,j \ldots . \end{cases}$$

Then each letter c_{ik} appears exactly twice in (5.4). Therefore (5.4) can be considered as the symbolic representation of a polyhedron P. It consists of six pentagons. They are mutually adjacent. For instance, country i is adjacent to country k along the edge c_{ik}. What we have to prove is that each vertex of P is of valence 3.

Consider any two successive numbers j, k in row i of scheme (5.3). Then in consequence of Rule R in (5.3) we find one of the four constellations:

$$\begin{array}{ll} i. \ \ldots j\,k \ldots & i. \ \ldots j\,k \ldots \\ j. \ \ldots i\,k \ldots & j. \ \ldots i\,k \ldots \\ k. \ \ldots i\,j \ldots & k. \ \ldots j\,i \ldots \end{array}$$

$$\begin{array}{ll} i. \ \ldots j\,k \ldots & i. \ \ldots j\,k \ldots \\ j. \ \ldots k\,i \ldots & j. \ \ldots k\,i \ldots \\ k. \ \ldots i\,j \ldots & k. \ \ldots j\,i \ldots . \end{array}$$

This means one of the four constellations

$$\begin{array}{ll} \ldots c_{ij} \ c_{ik} \ \ldots & \ldots c_{ij} \ c_{ik} \ \ldots \\ \ldots c_{ij} \ c_{jk} \ \ldots & \ldots c_{ij} \ c_{jk} \ \ldots \\ \ldots c_{ik}^{-1} \ c_{jk} \ \ldots & \ldots c_{jk}^{-1} \ c_{ik} \ \ldots \end{array}$$

$$\begin{array}{ll} \ldots c_{ij} \ c_{ik} \ \ldots & \ldots c_{ij} \ c_{ik} \ \ldots \\ \ldots c_{jk} \ c_{ij}^{-1} \ldots & \ldots c_{jk} \ c_{ij}^{-1} \ldots \\ \ldots c_{ik}^{-1} \ c_{jk}^{-1} \ldots & \ldots c_{jk} \ c_{ik} \ \ldots \end{array}$$

appears in (5.4). For all four, one can easily check that the vertex of country i incident with the two sides c_{ij} and c_{ik} is identified with exactly two other vertices. Fig. 5.1 illustrates the situation in the first case.

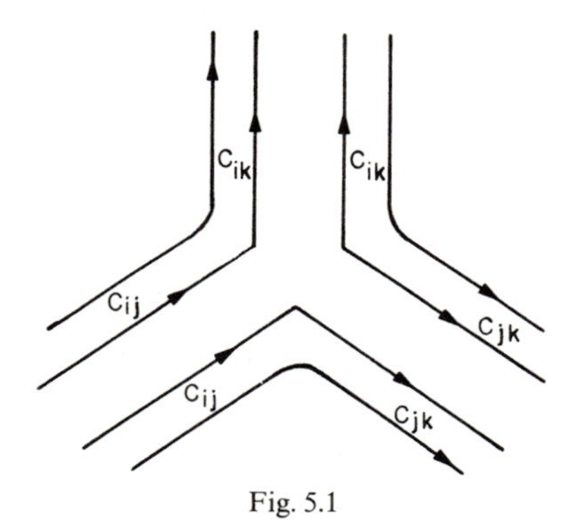

Fig. 5.1

The dual map of the polyhedron P is a triangular embedding of K_6 into a surface. Or, in general, P is a triangular embedding of K_n into a suitable surface.

As a first application of Theorem 5.2 consider the scheme

$$(5.5) \qquad \begin{array}{lllllllll} 0. & 1 & 4 & 7 & 5 & 6 & 8 & 3 & 2 \\ 3. & 4 & 7 & 1 & 8 & 0 & 2 & 6 & 5 \\ 6. & 7 & 1 & 4 & 2 & 3 & 5 & 0 & 8 \\ 1. & 8 & 5 & 2 & 0 & 4 & 6 & 7 & 3 \\ 4. & 2 & 8 & 5 & 3 & 7 & 0 & 1 & 6 \\ 7. & 5 & 2 & 8 & 6 & 1 & 3 & 4 & 0 \\ 2. & 6 & 3 & 0 & 1 & 5 & 7 & 8 & 4 \\ 5. & 0 & 6 & 3 & 4 & 8 & 1 & 2 & 7 \\ 8. & 3 & 0 & 6 & 7 & 2 & 4 & 5 & 1 \end{array}$$

which satisfies Rule R. The scheme (5.5) represents a triangular embedding of K_9 into a certain surface S that must be non-orientable because the case $n=9$ is not a regular one for orientable surfaces. By Theorem 4.4 we find the genus $q=5$, therefore $K_9 \lhd N_5$.

There is an obvious generalization of both Theorem 5.1 and 5.2: We can consider any graph G instead of the complete graph K_n. Let S be a surface.

Theorem 5.3. *If $G \lhd S$ then there exists a rotation of G such that the scheme of this rotation satisfies Rule R.*

Theorem 5.4. *If the scheme of a rotation of a graph G satisfies Rule R then there exists a suitable surface S such that $G \lhd S$.*

In order to prove both theorems we can transfer the proofs of Theorems 5.1 and 5.2 because in them we did not use the fact that K_n is the complete graph.

For instance the scheme

$$
\begin{array}{llllllllll}
0. & 3 & 1 & 7 & 2 & 5 & 6 & 9 & 4 & \\
1. & 4 & 2 & 6 & 8 & 3 & 0 & 7 & 5 & \\
2. & 5 & 3 & 6 & 1 & 4 & 7 & 0 & & \\
3. & 0 & 4 & 6 & 2 & 5 & 7 & 8 & 1 & \\
4. & 1 & 5 & 6 & 3 & 0 & 9 & 7 & 2 & \\
5. & 2 & 0 & 6 & 4 & 1 & 7 & 3 & & \\
6. & 1 & 2 & 3 & 4 & 5 & 0 & 9 & 7 & 8 \\
7. & 2 & 0 & 1 & 5 & 3 & 8 & 6 & 9 & 4 \\
8. & 3 & 1 & 6 & 7 & & & & & \\
9. & 4 & 0 & 6 & 7 & & & & &
\end{array}
\tag{5.6}
$$

satisfies Rule R. Therefore it represents a triangular embedding of a graph G into a certain surface S. Observe that the vertices $0, 1, \ldots, 7$ are mutually adjacent in G. Therefore G contains K_8 as a subgraph. For the graph G we easily find $\alpha_0=10$, $\alpha_1=36$. Theorem 4.4 says $\alpha_1 = 3\alpha_0 - 3E(S)$. Thus $E(S) = -2$. How can we find out whether S is orientable or not? The Euler characteristic does not give any information about that because it is an even number.

If we can reverse the cyclic order of the numbers in some of the rows in (5.6) such that Rule R^* of Section 2.2 holds, then S is orientable. (Compare with Fig. 4.8.) Otherwise it is not. Consider the local part

$$
\begin{array}{llllllllll}
1. & 4 & 2 & 6 & 8 & 3 & 0 & . & . & \\
2. & 5 & 3 & 6 & 1 & 4 & . & . & & \\
3. & . & . & 6 & 2 & 5 & . & 8 & 1 & 0
\end{array}
$$

of the scheme (5.6).

Whatever the chosen orientation of these three rows may be, the Rule R^* will not hold for *all* three pairs of rows. Therefore (5.6) represents a non-orientable surface and we have an embedding $K_8 \subset N_4$. Together with (4.18) this shows that $\bar{\gamma}(K_8)=4$.

If we specialize Theorem 5.3 to orientable surfaces we get the following

Theorem 5.5. *If $G \lhd S$ and S is an orientable surface, then there exists a triangular rotation of G.*

Recall that a rotation of G is called a triangular rotation whenever the scheme satisfies Rule R^*.

If we specialize Theorem 5.4 to triangular rotations we obtain

Theorem 5.6. *If there exists a triangular rotation of a graph G then there exists a triangular embedding of G into an orientable surface.*

If $n \equiv 7 \pmod{12}$ Theorem 2.8 says that there exists a triangular rotation of K_n. Therefore by Theorem 5.6 there exists an embedding $K_n \lhd S_p$ where p equals $(n-3)(n-4)/12$ (Theorem 4.4). Using (4.12) this proves Eq. (4.13) for all $n \equiv 7 \pmod{12}$.

For each other residue class (mod 12) we will prove (4.13) separately. Sometimes the method does not work for small values of n. We will consider these special cases first.

5.2. Orientable Special Cases

In this section we will determine the genus of the complete graph K_n for the special cases $n=8, 9, 10, 11, 12, 13, 14, 18, 20$, and 23. These are the cases (except 9, 10, 12) where the general method does not work. For the three cases 18, 20, 23 the solution was found by J. Mayer [54].

$n=8$. The scheme (2.9) does not satisfy rule R^*. But if we modify it to the form

$$
\begin{array}{llllllllll}
 & 0. & 2 & 7 & 3 & 1 & 4 & 5 & 6 & x \\
 & 2. & 4 & 1 & 5 & 3 & 6 & 7 & 0 & x \\
 & 4. & 6 & 3 & 7 & 5 & 0 & 1 & 2 & x \\
 & 6. & 0 & 5 & 1 & 7 & 2 & 3 & 4 & x \\
(5.7) & 1. & 7 & 6 & 5 & 2 & 4 & 0 & 3 & x \\
 & 3. & 1 & 0 & 7 & 4 & 6 & 2 & 5 & x \\
 & 5. & 3 & 2 & 1 & 6 & 0 & 4 & 7 & x \\
 & 7. & 5 & 4 & 3 & 0 & 2 & 6 & 1 & x \\
 & x. & (6 & 4 & 2 & 0) & (1 & 3 & 5 & 7)
\end{array}
$$

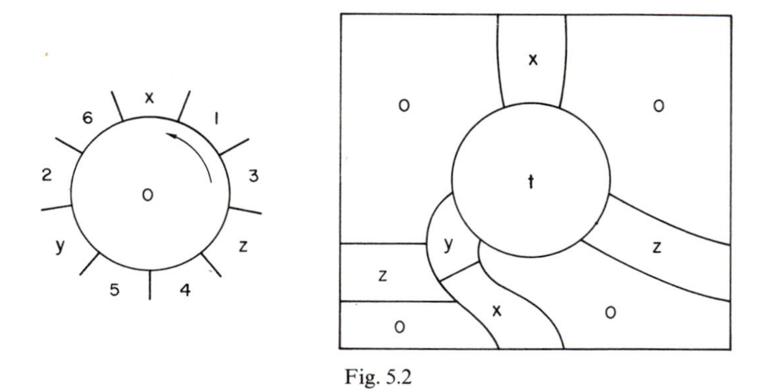

Fig. 5.2

rule R^* holds, with row x consisting of two cycles. In order to understand this scheme (5.7) replace row x by two rows x_0. 6420 and x_1. 1357, and the letter x by x_0 in rows 0, 2, 4, 6, and by x_1 in rows 1, 3, 5, 7. Then (5.7) represents a triangular rotation for a graph G which contains K_8 as a subgraph.

Theorem 5.6 says that G is triangularly embeddable into an orientable surface. Since $\alpha_0=10$ and $\alpha_1=4\cdot 7+8=36$ it follows from Theorem 4.4 that $p=2$. Therefore $K_8\subset S_2$. Together with (4.12) this proves $\gamma(K_8)=2$.

$n=9$. Use the same scheme (5.7) and consider the dual map. Excise the two quadrilaterals x_0 and x_1. The result is a surface with two holes in it. Identify the boundaries of the two holes with the boundary curves of an additional cylinder such that you obtain an orientable surface. This is exactly the surface S_3. Denote the extra cylinder by x. Then $1, 2, \ldots, 8, x$ are all adjacent to each other. Although x is not a country you can easily find an embedding of K_9 into S_3. Just proceed as in the proof of Theorem 4.6.

$n=10$. The scheme (2.8) of Chapter 2 presents a triangular embedding of the graph $K_{10}-K_3$ into an orientable surface S_p. Using Theorem 4.4 we obtain $p=3$. But formula (4.12) says $\gamma(K_{10})\geqq 4$ and we must prove that $\gamma(K_{10})=4$. This means we can only use one handle to connect the vertices x, y, z with each other by an arc. This we call the *additional adjacency problem.* We can handle it in the following way. Consider the dual map of the embedding $K_{10}-K_3\lhd S_3$. This map has only vertices of valence 3 and the countries x, y, z are not adjacent to each other. Fig. 5.2 illustrates the vicinity of country 0 as it is determined by the scheme (2.8). In the right part of Fig. 5.2 there is a certain map on a torus. Notice that 0 appears to be cut into four parts but is, in fact, a single country on the torus. Go to the surface at the left and excise the country 0. Do the same with country t of the torus. Identify the boundaries of the two resulting surfaces in the obvious way. After this there is a new country named 0 that is adjacent to the same countries as the old country 0 was before. Notice that we have gained the adjacencies between x, y and z.

So we have constructed a map with 10 mutually adjacent countries on the surface S_4. Hence $\gamma(K_{10})=4$.

$n=12s+10$. Entirely the same method works for the general case $n=12s+10$. In Chapter 2 we established a triangular rotation of the graph $K_{12s+10}-K_3$ (Theorem 2.9). This defines an embedding $K_{12s+10}-K_3 \lhd S_p$ where $\alpha_1=(6s+5)(12s+9)-3$ and $\alpha_0=12s+10$ hold. Using the formula $\alpha_1=3\alpha_0-3E(S)$ of Theorem 4.4, we obtain

$$(6s+5)(12s+9)-3=3(12s+10)-6+6p,$$

$$6p=72s^2+78s+21-3=(12s+7)(6s+3)-3,$$

$$12p=(n-3)(n-4)-6.$$

Thus p is one less than the right hand side of the inequality (4.12). So we have one handle at our disposal in order to manufacture the three missing adjacencies between x, y and z. This can be done exactly as in the previous special case $n=10$. This completes the proof of Eq. (4.13) for all $n\equiv 10 \pmod{12}$.

In the next cases we drop the computing of the genus p, because we have already had enough examples. Just use Theorem 4.4 and inequality (4.12).

$n=11$. The following scheme satisfies rule R^*. One can easily compute the genus p of the represented orientable surface. It is $p=5$.

1.	4	10	5	7	x	3	2	6	9	8	11	
2.	4	5	11	y	8	10	6	1	3	7	9	
3.	5	6	10	7	2	1	x	8	4	11	9	
4.	6	7	10	1	11	3	8	5	2	9		
5.	7	1	10	11	2	4	8	6	3	9		
6.	1	2	10	3	5	8	y	11	7	4	9	
7.	2	3	10	4	6	11	8	x	1	5	9	
8.	x	7	11	1	9	10	2	y	6	5	4	3
9.	3	11	10	8	1	6	4	2	7	5		
10.	5	1	4	7	3	6	2	8	9	11		
11.	9	3	4	1	8	7	6	y	2	5	10	
x.	1	7	8	3								
y.	2	11	6	8								

Since the first 11 vertices (or in dual form the first 11 countries) are mutually adjacent it follows $\gamma(K_{11})=5$.

$n=12$. This is a regular case. The scheme

0.	5	7	2	1	11	8	4	3	9	6	10
4.	9	11	6	5	3	0	8	7	1	10	2
8.	1	3	10	9	7	4	0	11	5	2	6
1.	8	6	11	0	2	5	9	10	4	7	3
5.	0	10	3	4	6	9	1	2	8	11	7
9.	4	2	7	8	10	1	5	6	0	3	11
2.	3	6	8	5	1	0	7	9	4	10	11
6.	7	10	0	9	5	4	11	1	8	2	3
10.	11	2	4	1	9	8	3	5	0	6	7
3.	2	11	9	0	4	5	10	8	1	7	6
7.	6	3	1	4	8	9	2	0	5	11	10
11.	10	7	5	8	0	1	6	4	9	3	2

represents a triangular embedding of K_{12} into S_6.

$n=13$. The scheme

1.	2	3	4	5	6	7	8	9	10	11	12	13	
2.	5	3	1	13	7	11	8	12	9	x	4	10	6
3.	4	1	2	5	11	10	7	13	9	6	12	8	
4.	1	3	8	6	9	12	7	10	2	x	11	13	5
5.	11	3	2	6	1	4	13	12	10	8	7	9	
6.	12	3	9	4	8	13	11	7	1	5	2	10	
7.	4	12	x	9	5	8	1	6	11	2	13	3	10
8.	2	11	9	1	7	5	10	13	6	4	3	12	
9.	10	1	8	11	5	7	x	2	12	4	6	3	13
10.	3	11	1	9	13	8	5	12	6	2	4	7	
11.	10	3	5	9	8	2	7	6	13	4	x	12	1
12.	7	4	9	2	8	3	6	10	5	13	1	11	x
13.	6	8	10	9	3	7	2	1	12	5	4	11	
x.	7	12	11	4	2	9							

represents a triangular embedding of a graph into S_8. The graph contains K_{13} as a subgraph. Therefore $\gamma(K_{13})=8$.

$n=14$. The scheme

0.	6	2	7	5	10	4	1	y	3	11	x	9	8
2.	8	4	9	7	0	6	3	y	5	1	x	11	10
4.	10	6	11	9	2	8	5	y	7	3	x	1	0
6.	0	8	1	11	4	10	7	y	9	5	x	3	2
8.	2	10	3	1	6	0	9	y	11	7	x	5	4
10.	4	0	5	3	8	2	11	y	1	9	x	7	6
1.	7	11	6	8	3	9	10	y	0	4	x	2	5
3.	9	1	8	10	5	11	0	y	2	6	x	4	7
5.	11	3	10	0	7	1	2	y	4	8	x	6	9
7.	1	5	0	2	9	3	4	y	6	10	x	8	11
9.	3	7	2	4	11	5	6	y	8	0	x	10	1
11.	5	9	4	6	1	7	8	y	10	2	x	0	3
x.	2	1	4	3	6	5	8	7	10	9	0	11	
y.	10	11	8	9	6	7	4	5	2	3	0	1	

represents a triangular rotation of the graph $K_{14}-K_2$ or in dual form a map with 14 countries. Only the countries x and y are not adjacent. The genus of the map is 9. With one extra handle you can produce the adjacency of x with y. Therefore $\gamma(K_{14})\leqq 10$. From (4.12) we know that the reverse inequality is also true. Hence $\gamma(K_{14})=10$. In Chapter 9 we will describe how to get this scheme for $n=14$ by using a current graph.

$n=18$, 20 and 23. These three special cases are unfortunately very irregular. They possess no symmetries at all. In the scheme for $n=18$ the row 5 appears as two cycles. As in the special case $n=9$ we can interpret the scheme as a map having 17 countries and one cylinder. All 18 are mutually adjacent.

$n=18$ $p=18$

1.	14	17	9	18	7	5	12	4	13	10	6	2	11	8	3	16	15
2.	1	6	9	8	16	12	10	5	13	14	7	17	18	4	15	3	11
3.	1	8	4	18	9	7	14	5	6	11	2	15	10	17	13	12	16
4.	1	12	15	2	18	3	8	10	9	5	7	11	17	14	6	16	13
5.	(4	9	12	1	7)	(6	3	14	15	13	2	10	8	17	11	16	18)
6.	1	10	16	4	14	11	3	5	18	8	13	17	12	7	15	9	2
7.	1	18	16	17	2	14	3	9	10	13	15	6	12	8	11	4	5
8.	1	11	7	12	14	16	2	9	13	6	18	15	17	5	10	4	3
9.	1	17	16	14	13	8	2	6	15	11	12	5	4	10	7	3	18
10.	1	13	7	9	4	8	5	2	12	11	14	18	17	3	15	16	6
11.	1	2	3	6	14	10	12	9	15	18	13	16	5	17	4	7	8
12.	1	5	9	11	10	2	16	3	13	18	14	8	7	6	17	15	4
13.	1	4	16	11	18	12	3	17	6	8	9	14	2	5	15	7	10
14.	1	15	5	3	7	2	13	9	16	8	12	18	10	11	6	4	17
15.	1	16	10	3	2	4	12	17	8	18	11	9	6	7	13	5	14
16.	1	3	12	2	8	14	9	17	7	18	5	11	13	4	6	10	15
17.	1	14	4	11	5	8	15	12	6	13	3	10	18	2	7	16	9
18.	1	9	3	4	2	17	10	14	12	13	11	15	8	6	5	16	7

$n=20$ $p=23$

1.	8	4	5	17	11	16	15	19	6	14	9	3	20	2	12	13	7	10	18	
2.	1	20	9	11	5	13	3	16	19	7	15	6	4	14	18	10	8	17	12	
3.	1	9	4	19	5	7	11	17	15	10	12	16	2	13	14	6	8	18	20	
4.	1	8	14	2	6	13	17	18	11	12	19	3	9	20	10	15	16	7	5	
5.	1	4	7	3	19	10	9	14	16	20	13	2	11	x	8	15	12	6	18	17
6.	1	19	17	10	16	13	4	2	15	9	18	5	12	7	20	11	8	3	14	
7.	1	13	15	2	19	18	14	17	20	6	12	8	11	3	5	4	16	9	10	
8.	1	18	3	6	11	7	12	20	17	2	10	13	9	15	5	x	19	16	14	4
9.	1	14	5	10	7	16	17	19	12	18	6	15	8	13	11	2	20	4	3	
10.	1	7	9	5	19	11	14	12	3	15	4	20	16	6	17	13	8	2	18	
11.	1	17	3	7	8	6	20	15	14	10	19	x	5	2	9	13	12	4	18	16
12.	1	2	17	16	3	10	14	20	8	7	6	5	15	18	9	19	4	11	13	
13.	1	12	11	9	8	10	17	4	6	16	18	19	14	3	2	5	20	y	15	7
14.	1	6	3	13	19	20	12	10	11	15	17	7	18	2	4	8	16	5	9	
15.	1	16	4	10	3	17	14	11	20	18	12	5	8	9	6	2	7	13	y	19
16.	1	11	18	13	6	10	20	5	14	8	19	2	3	12	17	9	7	4	15	
17.	1	5	18	4	13	10	6	19	9	16	12	2	8	20	7	14	15	3	11	
18.	1	10	2	14	7	19	13	16	11	4	17	5	6	9	12	15	20	3	8	
19.	1	15	y	20	14	13	18	7	2	16	8	x	11	10	5	3	4	12	9	17	6
20.	1	3	18	15	11	6	7	17	8	12	14	19	y	13	5	16	10	4	9	2
x.	5	11	19	8																
y.	15	13	20	19																

$n=23$ $p=32$

1.	23	19	12	17	6	9	2	7	18	20	8	5	16	14	3	11	22	21	15	13	4	10	
2.	1	9	20	15	4	11	5	13	3	16	19	6	21	22	17	14	10	8	18	23	12	7	
3.	1	14	23	5	17	15	10	22	16	2	13	18	6	8	20	9	19	4	12	21	7	11	
4.	1	13	22	18	9	11	2	15	23	6	16	8	7	14	17	21	20	5	12	3	19	10	
5.	1	8	12	4	20	6	15	18	22	19	17	3	23	7	21	9	14	13	2	11	10	x	16
6.	1	17	10	16	4	23	13	21	2	19	15	5	20	12	11	7	22	y	8	3	18	14	9
7.	1	2	12	13	15	19	14	4	8	17	20	22	6	11	3	21	5	23	16	9	10	18	
8.	1	20	3	6	y	23	14	22	9	17	7	4	16	21	18	2	10	13	19	11	15	12	5
9.	1	6	14	5	21	13	17	8	22	12	23	10	7	16	15	11	4	18	19	3	20	2	
10.	1	4	19	x	5	11	21	12	22	3	15	20	16	6	17	13	8	2	14	18	7	9	23
11.	1	3	7	6	12	14	21	10	5	2	4	9	15	8	19	18	17	16	13	20	23	22	
12.	1	19	13	7	2	23	9	22	10	21	3	4	5	8	15	14	11	6	20	18	16	17	
13.	1	15	7	12	19	8	10	17	9	21	6	23	18	3	2	5	14	20	11	16	22	4	
14.	1	16	20	13	5	9	6	18	10	2	17	4	7	19	21	11	12	15	22	8	23	3	
15.	1	21	23	4	2	20	10	3	17	22	14	12	8	11	9	16	18	5	6	19	7	13	
16.	1	5	x	19	2	3	22	13	11	17	12	18	15	9	7	23	21	8	4	6	10	20	14
17.	1	12	16	11	18	21	4	14	2	22	15	3	5	19	23	20	7	8	9	13	10	6	
18.	1	7	10	14	6	3	13	23	2	8	21	17	11	19	9	4	22	5	15	16	12	20	
19.	1	23	17	5	22	20	21	14	7	15	6	2	16	x	10	4	3	9	18	11	8	13	12
20.	1	18	12	6	5	4	21	19	22	7	17	23	11	13	14	16	10	15	2	9	3	8	
21.	1	22	2	6	13	9	5	7	3	12	10	11	14	19	20	4	17	18	8	16	23	15	
22.	1	11	23	y	6	7	20	19	5	18	4	13	16	3	10	12	9	8	14	15	17	2	21
23.	1	10	9	12	2	18	13	6	4	15	21	16	7	5	3	14	8	y	22	11	20	17	19
x.	5	10	19	16																			
y.	6	22	23	8																			

5.3. Outline for General Cases

In order to determine the genus of the complete graph K_n we have to find a certain embedding of K_n into an orientable surface of minimal genus. If $n \equiv 0$, 3, 4 or 7 (mod 12) this embedding is expected to be triangular. For these regular cases we have already constructed (Chapter 2), or we will construct (Chapters 6, 9, 11), such triangular embeddings of K_n into an orientable surface.

But what should we do in the other cases? We have good methods for constructing triangular rotations (or orientable embeddings) of graphs. But we know that for the non-regular cases there is no triangular rotation of K_n. As we have seen in cases $n=10$ and $n=14$ we first constructed an embedding of a subgraph of K_n. Then we were able to create the missing adjacencies by using one additional handle. Let us discuss what possibilities we have for applying similar methods in the other non-regular cases.

Assume there exists a triangular embedding of a graph, obtained from K_n by removing t arcs, into an orientable surface S_p.

Let us write this situation symbolically:

$$K_n - t \text{ arcs} \lhd S_p .$$

Using the formula $\alpha_1 = 3\alpha_0 - 6 + 6p$ of Theorem 4.4, we obtain

$$n(n-1) - 2t = 6n + 12p - 12,$$

$$(n-3)(n-4) \equiv 2t \pmod{12}.$$

This gives us four groups of cases:

a) If $n \equiv 0$, 3, 4 or 7 (mod 12) then $t \equiv 0 \pmod 6$.

b) If $n \equiv 2$ or 5 (mod 12) then $t \equiv 1 \pmod 6$.

c) If $n \equiv 1$, 6, 9 or 10 (mod 12) then $t \equiv 3 \pmod 6$.

d) If $n \equiv 8$ or 11 (mod 12) then $t \equiv 4 \pmod 6$.

In the two cases b) the natural approach is to construct a triangular embedding of $K_n - K_2$. We will be able to do this for $n \equiv 5 \pmod{12}$ in Chapter 9. The missing adjacency can be made easily by one handle. For $n \equiv 2 \pmod{12}$ an orientable triangular embedding of $K_n - K_2$ was recently found by Jungerman [41]. However his solution is difficult. In this book we will use a very adventurous method to determine the genus $\gamma(K_n)$ for $n \equiv 2 \pmod{12}$. The method is much easier to understand than to discover.

In the four cases c) we can expect the existence of an orientable triangular embedding of $K_n - K_3$. For $n \equiv 10 \pmod{12}$ we have already constructed such an embedding of $K_n - K_3$. We will do the same in the cases $n \equiv 1$ and 6 (mod 12) in Chapters 6 and 9. For $n \equiv 9 \pmod{12}$ orientable triangular embeddings of $K_n - K_3$ and even $K_n - K_6$ have been found by Youngs [102] and Guy and Ringel [33]. In both cases the extra adjacency problem can be solved. But in this book we present an elegant solution using another method.

In the two cases d) the natural approach is to find an orientable triangular embedding of $K_n - K_5$, and then make the ten missing adjacencies with two additional handles. We will do this for all $n \equiv 11 \pmod{12}$ in Chapter 7. In the case $n \equiv 8 \pmod{12}$ we were able to find a triangular rotation of $K_n - K_5$. However we could not solve the additional adjacency part of the problem. So we will solve this case another way.

In order to avoid a repeated calculation of the genus of a given triangular embedding in the next chapters, we prove

Theorem 5.7. *Suppose we have a triangular embedding of a graph of the form ($K_n - t$ arcs) into an orientable surface. If we are able to manufacture the t missing adjacencies by using exactly $\{t/6\}$ additional handles then the Eq.* (4.13) *is true for that integer n.*

Proof. By assumption we have the two embeddings:

$$K_n - t \text{ arcs} \lhd S_p,$$

$$K_n \subset S_{p+\{t/6\}}.$$

Applying the formula $\alpha_1 = 3\alpha_0 - 6 + 6p$ of Theorem 4.4 to the first embedding we obtain

$$p = \frac{(n-3)(n-4) - 2t}{12}.$$

From the second embedding we get

$$\gamma(K_n) \leqq p + \{t/6\} = \frac{(n-3)(n-4) - 2t}{12} + \left\{\frac{2t}{12}\right\}$$

$$= \left\{\frac{(n-3)(n-4)}{12}\right\}.$$

By (4.12) the above inequality really holds as an equality. This completes the proof of Theorem 5.7.

The general outline for the determination of the non-orientable genus of K_n is much simpler. For all $n \equiv 0$ or 1 (mod 3), $6 \leqq n \neq 7$, we will be able to show that a triangular embedding of K_n into a non-orientable surface N_q exists. Using the formula $\alpha_1 = 3\alpha_0 - 6 + 3q$ of Theorem 4.4 we then obtain $q = (n-3)(n-4)/6$.

For all $n \equiv 2$ (mod 3), $n \geqq 1$, the natural approach is to construct a triangular embedding of $K_n - K_2$ into a surface S. The scheme

(5.8)

0.	4	1	x	8	3	5	6	2	y	7
1.	5	2	x	0	4	6	7	3	y	8
2.	6	3	x	1	5	7	8	4	y	0
3.	7	4	x	2	6	8	0	5	y	1
4.	8	5	x	3	7	0	1	6	y	2
5.	0	6	x	4	8	1	2	7	y	3
6.	1	7	x	5	0	2	3	8	y	4
7.	2	8	x	6	1	3	4	0	y	5
8.	3	0	x	7	2	4	5	1	y	6
x.	0	1	2	3	4	5	6	7	8	
y.	0	2	4	6	8	1	3	5	7	

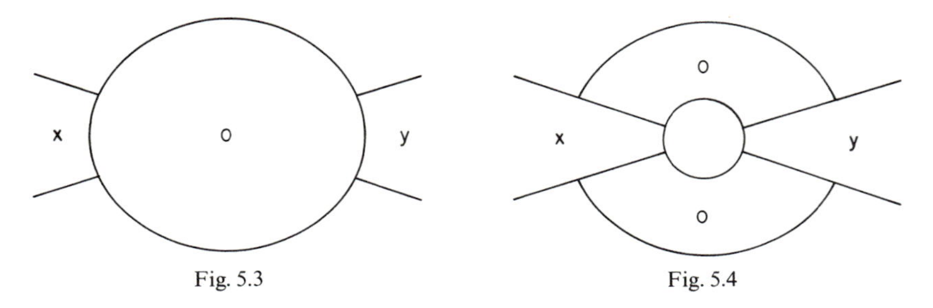

Fig. 5.3 Fig. 5.4

satisfies Rule R. Therefore it represents an example of such an embedding if $n=11$. Only the two vertices (or in dual form the two countries) x and y are not adjacent.

For all n with $n \equiv 2$, 5, or 11 (mod 12) we will construct an embedding $K_n - K_2 \lhd S$ later. It is not necessary to know whether S is orientable or not. Using one additional cross cap it is easy to perform the missing adjacency in the following way (the cross cap creates a non-orientable surface S' anyway).

Consider the map on S dual to the embedding $K_n - K_2 \lhd S$. Let x and y be the two non-adjacent countries. Each other country is adjacent to x and y. Let 0 be one of those countries. The situation around country 0 is shown in Fig. 5.3.

Excise the interior of a closed 2-cell in country 0 and identify opposite points on the boundary. This construction adds a cross cap to S giving us S'. The countries x and y are enlarged (appropriating some of the territory in country 0) so as to become adjacent to each other as in Fig. 5.4.

Notice that country 0 is still connected and has lost none of its adjacencies. In fact the adjacencies $(0, x)$ and $(0, y)$ each occur twice. The result is a map on S' consisting of n countries each adjacent to all the others. The dual of this decomposition of S' has a 1-skeleton containing K_n.

Using the formula $\alpha_1 = 3\alpha_0 - 3E(S)$ of Theorem 4.4 a direct computation shows that

$$E(S) = 3 - \left\{ \frac{(n-3)(n-4)}{6} \right\}.$$

Since we obtained S' by adding a cross cap, $E(S') = E(S) - 1$. Therefore we have found an embedding $K_n \subset S'$ where S' is the non-orientable surface of genus $\{(n-3)(n-4)/6\}$.

If $n \equiv 8$ (mod 12) and $n \geqq 20$ it is probably very difficult to find an embedding $K_n - K_2 \lhd S$. It has been proved that a triangular embedding of $K_8 - K_2$ into a (orientable or non-orientable) surface does not exist (see Ringel [68]). However we will use another method to determine the non-orientable genus of K_n in the case $n \equiv 8$ (mod 12).

6. Orientable Cases 1, 4, and 9

6.1. Orientable Case 4

Now we will construct a triangular embedding of K_n into an orientable surface for each n of the form $n=12s+4$.

If one tries the same method as in Case 7 in Section 2.3, employing the cyclic group $\mathbb{Z}_{12s+4}$, it will not succeed. Instead we will use the direct product of two cyclic groups, the group $\Gamma=\mathbb{Z}_2\times\mathbb{Z}_{6s+2}$ which consists of all pairs (a, b) where a is an element of $\mathbb{Z}_2$ and b an element of $\mathbb{Z}_{6s+2}$. The addition (group operation) is defined by

$$(a, b)+(a', b')=(a+a', b+b').$$

The identity element is $(0, 0)$. In the element (a, b) we call a the first and b the second *component*.

First consider the following solution for the smallest case $n=4$ using the group $\mathbb{Z}_2\times\mathbb{Z}_2$:

$$\begin{array}{llll}
(0,0). & (1,0) & (0,1) & (1,1)\\
(0,1). & (1,1) & (0,0) & (1,0)\\
(1,0). & (0,0) & (1,1) & (0,1)\\
(1,1). & (0,1) & (1,0) & (0,0)
\end{array}$$

Notice that rule R^* holds and that the row $(0, 0)$ generates the other rows by using the additive rule. This scheme describes the tetrahedron on the sphere which is a triangular embedding of K_4.

Let us now describe the next case $n=16$. We use the current graph of Fig. 6.1. It is convenient to record only the second component of the current, the first component being represented by a heavily drawn arc if it is 1, and a lightly drawn one if it is 0. The currents α of each of the three dead-end-arcs have the property $\alpha=-\alpha$, which means α is of order 2 in the group.

The construction principles C1), C2), C3), C4) of Section 2.3 are valid with respect to the group $\mathbb{Z}_2\times\mathbb{Z}_8$. In addition we use a new construction principle, namely,

C6) *Each element of order 2 in the group appears on a dead-end-arc.*

When we run through the graph along the circuit induced by the given rotation we should imagine that we turn at the middle of the dead-end-arc carrying current α. This is reasonable because $\alpha = -\alpha$ and we do not want to record any current twice (see Fig. 6.1). Thus in writing down row (0, 0), we record the element α just once. We have to prove that doing this leads to a scheme for K_{16}, and in general for K_{12s+4}, satisfying Rule Δ^*.

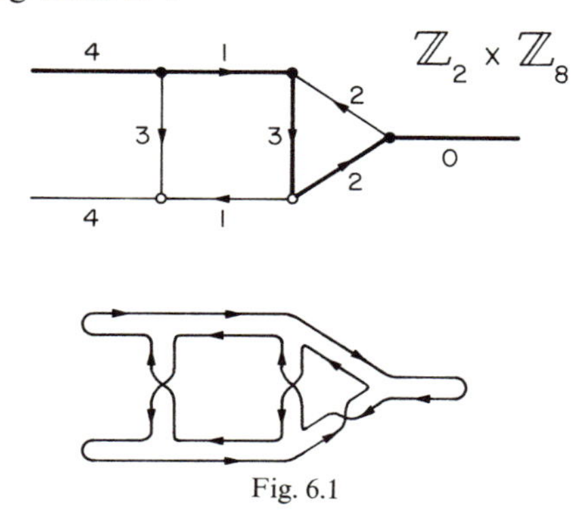

Fig. 6.1

Let i, j, k be elements of the group used (in this case $\mathbb{Z}_2 \times \mathbb{Z}_8$) and 0 the identity element. Assume we find in row i,

$$i. \ldots j\, k \ldots .$$

Then the additive rule says that

$$0. \ldots j-i\;\; k-i \ldots$$

appears in row 0. This is part of the log of the current graph with the local part shown in Fig. 2.16. Consequently

$$0. \ldots j-i, k-i \ldots i-k, h \ldots$$

appears in row 0 if $k-i$ is not of order 2, and

$$0. \ldots j-i, k-i, h \ldots$$

appears if $k-i$ is of order 2 ($k-i=i-k$). In any event we have

$$k. \ldots i\;\; k+h \ldots$$

in the row k. Kirchhoff's Current Law (Property C4) says that $j=k+h$. Thus rule Δ^* is proved.

In the general case $n=12s+4$, we use the same ladder-like graph (Fig. 2.19) which we already used in Case 10. There are 3 dead-end-arcs and $2s$ vertical arcs called rungs. The rotation on the top (on the bottom) of each rung is clockwise (counterclockwise). This guarantees one single circulation. Therefore construction principle C2) is satisfied. In the group $\mathbb{Z}_2 \times \mathbb{Z}_{6s+2}$ there are three elements of order 2, namely, $(1, 3s+1)$, $(0, 3s+1)$, and $(1, 0)$. We assign the element $(1, 3s+1)$ to the dead-end-arc at the top and let $(0, 3s+1)$ be the dead-end-arc below it. Assign $(1, 0)$ to the dead-end-arc at the right. If s is odd we choose the second components of the currents on the rungs to be

$$\overline{3s}, 3s-3, \overline{3s-6} \dots 6, \bar{3}, \overline{3s}, \bar{3}, 6, \bar{9} \dots 3s-3$$

in that order. The bars mean that the orientation of that rung goes down, otherwise it points up. All the remaining currents and directions are completely determined by Kirchhoff's Current Law (C4). This leads to the sequence (second components)

$$1, 3s-2, 4, 3s-5, 7 \dots$$

for the first s horizontals on top and on the lower part. For the other half of the horizontals one gets the sequence (second component)

$$\dots 8, 3s-4, 5, 3s-1, 2.$$

In Fig. 6.2 the cases $s=3$ and $s=5$ are shown.

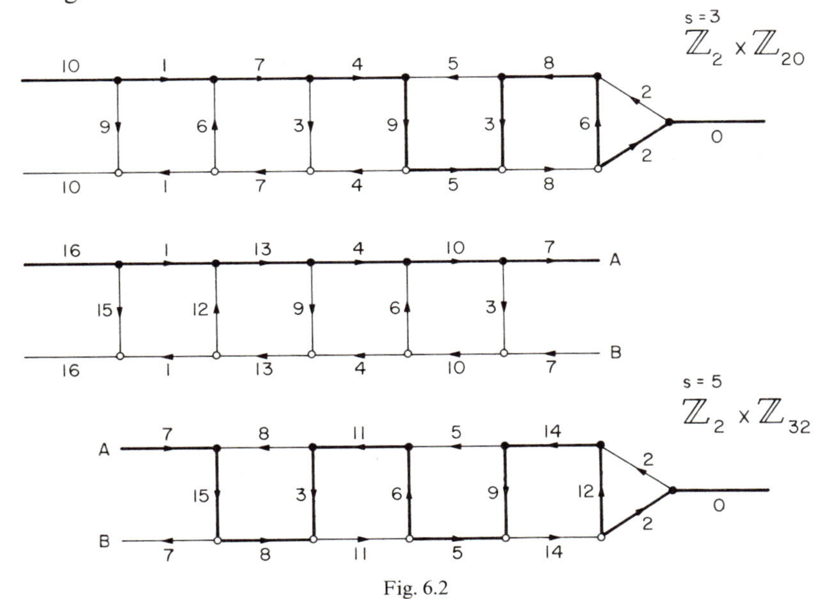

Fig. 6.2

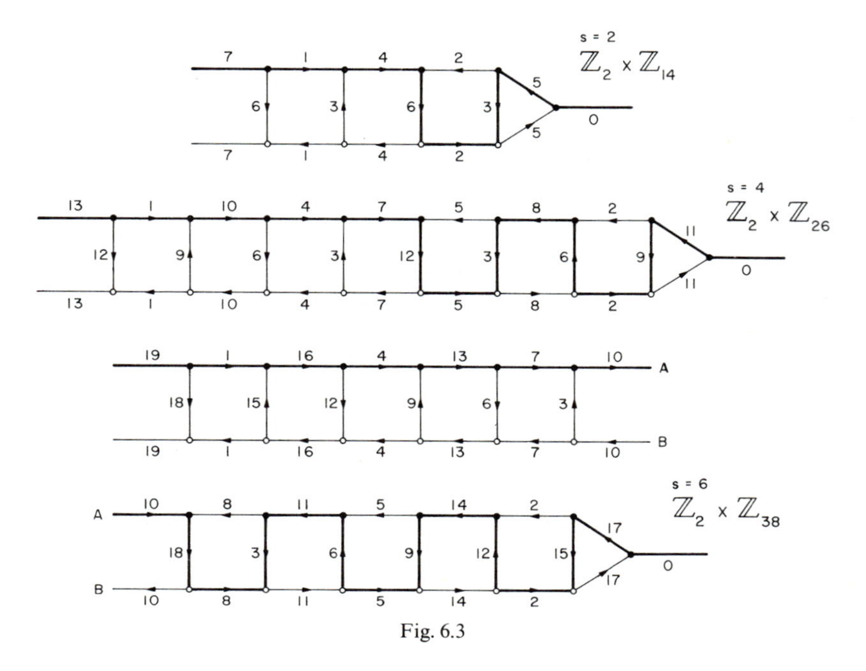

Fig. 6.3

Now we display the first component of the currents in the following manner. Just find a way from the top left horizontal to the last one on the right and use each number (second component) exactly once. Then draw all arcs of this way heavily which means the first component is 1. This guarantees KCL for the first components also. Of course there are several methods of doing this. (See Fig. 6.2.)

The solution, if s is even, is similar. Choose for the rungs the currents (second components)

$$\overline{3s}, 3s-3, \overline{3s-6} \ldots 9, \overline{6}, 3, \overline{3s}, \overline{3}, 6, \overline{9} \ldots \overline{3s-3}.$$

The first small cases are shown in Fig. 6.3.

6.2. Arithmetic Combs

A great help in distributing the elements or numbers on the rungs of a ladder-like current graph is the method of using an arithmetic sequence and alternating directions of the rungs. In Fig. 6.4 partial pictures from Fig. 2.15 (top) and Fig. 6.3 (bottom) are shown. In the latter partial picture the order has been reversed. Both partial pictures have the form of a comb. It will be called an *arithmetic comb* if

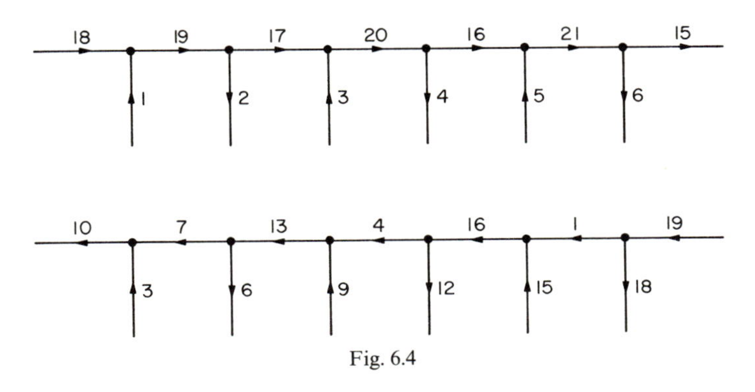

Fig. 6.4

a) the rungs (or teeth) alternate direction and the currents form an arithmetic sequence and

b) the currents on the horizontals are such that Kirchhoff's Current Law from Section 2.3 holds at each vertex of valence three.

We will use arithmetic combs very often, in fact, as often as possible. In some of the cases, however, we have to use combs satisfying only condition b) and not a).

When we use the expression *comb* we always mean that condition b) is satisfied. Normally one has to mention from what group the currents come. In the two examples of Fig. 6.4, this does not matter because KCL holds anyway (even in the ring $\mathbb{Z}$ of all integers).

It is convenient to describe a comb in the form of two rows, the first coming from the currents on the horizontals, the second from the currents on the teeth. So the notation for the first comb of Fig. 6.4 is

$$\begin{matrix} 18 & & 19 & & 17 & & 20 & & 16 & & 21 & & 15 \\ & 1 & & 2 & & 3 & & 4 & & 5 & & 6 & \end{matrix}$$

It is not necessary to mention how the arcs are oriented because this is uniquely determined by KCL if the orientation of *one* arc is given.

6.3. Orientable Case 1

The Euler formula suggests (Section 5.3) that in the case $n=12s+1$ one can expect the existence of a triangular embedding of the graph K_n-K_3 into an orientable surface. For the smallest case $n=13$ it is still unknown whether an embedding of $K_{13}-K_3$ exists or not. But for all $s\geqq 2$ we now present a constructive existence proof. There is a solution which is very simple and pleasing but it has the disadvantage of working only for even s. We present it first.

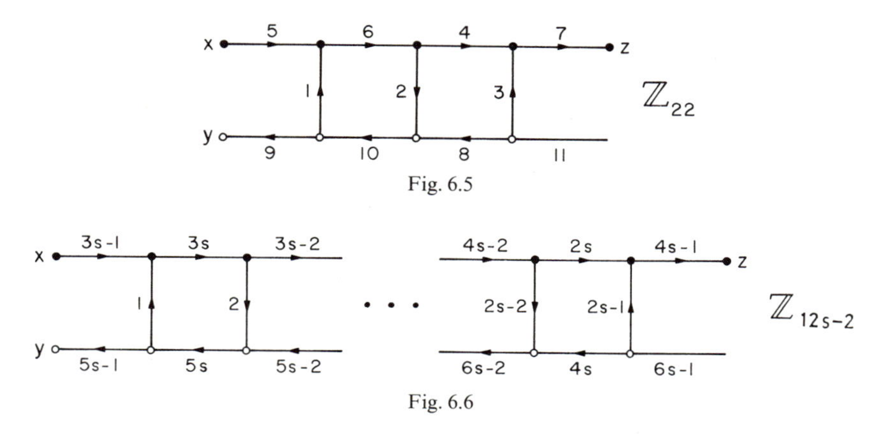

Fig. 6.5

Fig. 6.6

We use the letters x, y, and z to identify the exceptional vertices of $K_{12s+1}-K_3$ and the elements of the group $\mathbb{Z}_{12s-2}$ for the other vertices.

Consider the current graph of Fig. 6.5 for $s=2$ and of Fig. 6.6 for general s. One can see it is based on a simple arithmetic comb. Each element $1, 2, \ldots, 6s-1$ of $\mathbb{Z}_{12s-2}$ appears exactly once in the current graph. The only element of order 2, namely $6s-1$, is found on a dead-end-arc. The given rotation induces a single circuit. Kirchhoff's Current Law holds at each vertex of valence 3. The currents on the three dead-end-arcs at x, y, z are $3s-1$, $5s-1$, and $4s-1$. We compute the greatest common divisor these have with $12s-2$.

$$(3s-1, 12s-2)=(3s-1, 2)=(s-1, 2)$$

$$(5s-1, 12s-2)=(5s-1, 2s)=(s-1, 2s)=(s-1, 2)$$

$$(4s-1, 12s-2)=(4s-1, 1)=1.$$

The first two equal 1 only if *s is even*. Therefore all the properties C1) to C6) from Sections 2.3 and 6.1 are satisfied.

Just as in Case 10 this produces a row 0 and by the Additive Rule of Section 2.3 a triangular rotation for the graph $K_{12s+1}-K_3$. We also say this is a scheme for $K_{12s+1}-K_3$ satisfying Rule Δ^*. This solves the embedding problem if s is even.

Now we describe a construction which works for all $s \geqq 2$. It was discovered by Gustin in 1964 who unfortunately did not publish it. The vertices are identified by x, y, z and the elements of the group $\mathbb{Z}_2 \times \mathbb{Z}_{6s-1}$. It should be mentioned that this direct product is really isomorphic to $\mathbb{Z}_{12s-2}$ because $(2, 6s-1)=1$. But we do not use this fact. By the way, the direct product we used in Section 6.1 was not a cyclic group.

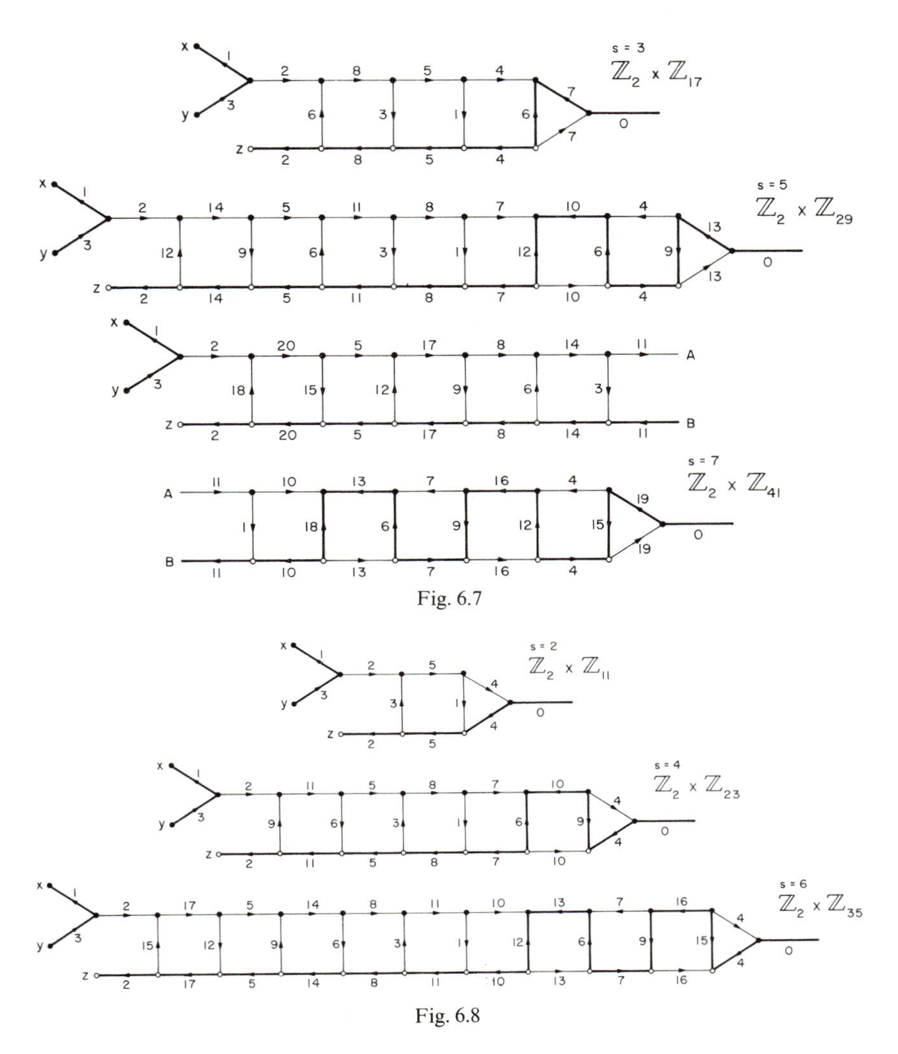

Fig. 6.7

Fig. 6.8

We will use all six construction principles C 1) to C 6) from Sections 2.3 and 6.1. Consider the ladder-like current graph of Fig. 6.7 in cases $s=3$, 5, or 7, and of Fig. 6.8 in the even cases $s=2$, 4, or 6. We will describe the general pattern soon. Arcs carrying currents with first component 1 are drawn heavily.

Notice that the currents on the dead-end-arcs are (1, 1), (1, 3), (1, 2), and (1, 0). The first three are generators of the group and produce the rows x, y, z (C 5). The last, (1, 0), is the only element of order two, which is in harmony with the construction principle C 6). There is an even number of rungs and therefore there exists a rotation inducing just one single circuit. Further, it contains each non-trivial element of the group exactly once.

At each vertex of valence three, Kirchhoff's Current Law holds. For the first component KCL is easy to check because at each vertex of valence three there are either two heavily drawn arcs or none. So we only have to check KCL for the second component.

We have to explain what the pattern is for general values of s. The distribution of the first component is quite obvious, so we describe the distribution of the second component, which consists of the nonzero elements of the group $\mathbb{Z}_{6s-1}$. The first horizontal on the top is directed to the right and carries the current 2. The two adjacent dead-end-arcs carry 1 and 3. There are $2s-2$ verticals. The first half of them form a simple arithmetic comb having multiples of 3 on the rungs. The horizontals of this comb exhaust all the currents $\equiv 2 \pmod 3$. The next vertical carries the current 1 and is directed downwards. On all the remaining verticals are multiples of 3. This forces the horizontals of the right half to have currents $\equiv 1 \pmod 3$.

If s is an odd number the currents on the verticals are

$$3s-3 \dots 12, \bar{9}, 6, \bar{3}, \bar{1}, 3s-3, 6, \bar{9}, 12 \dots \overline{3s-6}.$$

The bar means that the orientation on the rung points down.

Not only do the first $s-1$ verticals generate an arithmetic comb but the last $s-3$ verticals do too. Therefore it is easy to see that all the numbers $1, 2, \dots, 3s-1$ appear exactly twice, once on a heavily drawn arc and once on a light one. The number 0 is used just once, namely on a heavy dead-end-arc.

If s is even the sequence of the currents for the verticals is

$$3s-3, \dots, \overline{12}, 9, \bar{6}, 3, \bar{1}, 3s-6, 6, \bar{9}, 12, \dots, \overline{3s-9}, \overline{3s-3}.$$

The first arithmetic comb is almost the same as in the odd case. The second arithmetic comb begins at the same place with $6, \bar{9}, 12, \dots$ but it runs only up to $\overline{3s-9}$.

One might ask why the present method does not work for the smallest case $s=1$. The answer is: the group $\mathbb{Z}_2 \times \mathbb{Z}_5$ which is equivalent to the group $\mathbb{Z}_{10}$ does not have enough generators.

6.4. Coil Diagrams

We would like to explain a method which helps to illustrate solutions such as the ones given in this chapter. In fact it is the method by which the solutions were discovered.

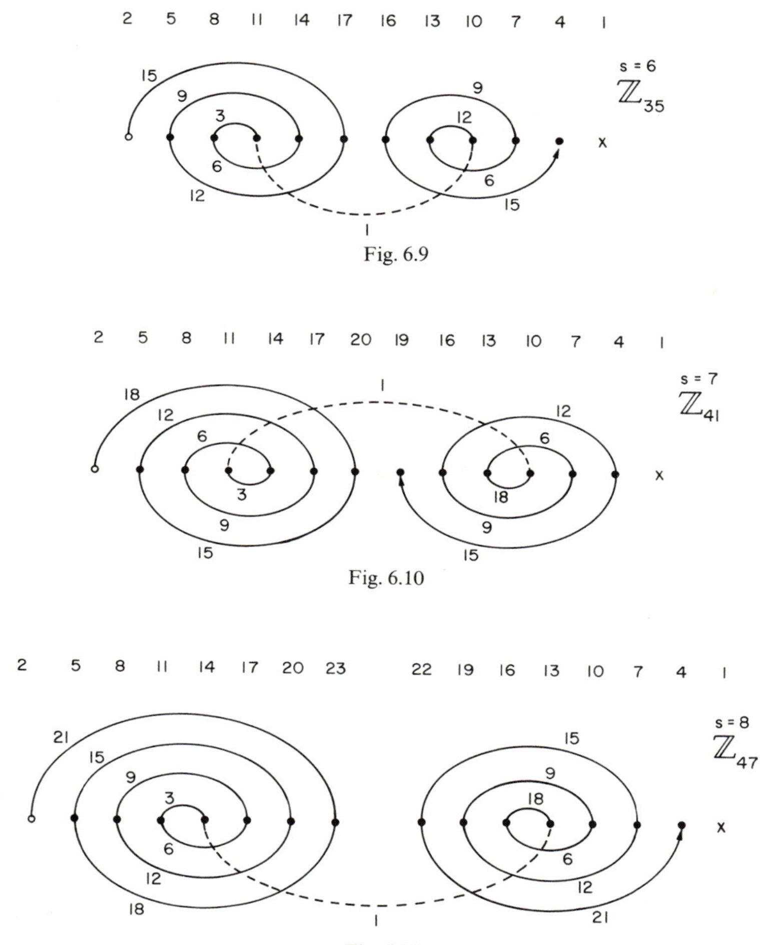

Fig. 6.9

Fig. 6.10

Fig. 6.11

Consider the picture for $s=6$ in Fig. 6.8. On the first half of the horizontals there is a permutation of the numbers 2, 5, 8, ..., 17 and on the second half a permutation of the numbers 4, 7, ..., 16. We present these numbers on a straight line as equidistant points marked with dots. (See Fig. 6.9.) We draw an arch from dot 2 to dot 17, then from dot 17 to dot 5 and so on. In other words we draw arches beginning with dot 2 such that we obtain a *coil* (i.e., a sequence of arches) passing through the dots in exactly the same order as the numbers appear on the horizontals in Fig. 6.8 ($s=6$). Then assign to each arch the current f on the vertical which is between the two horizontals c and d in Fig. 6.8 ($s=6$) corresponding to the two dots c and d of the arch. This is illustrated in Fig. 6.12.

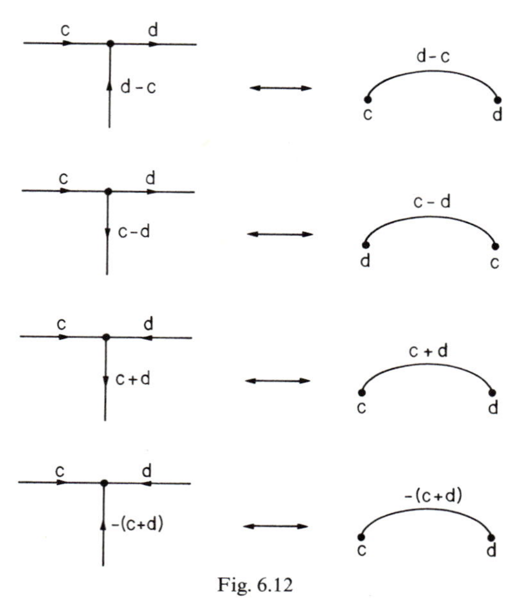

Fig. 6.12

Since in the current graph KCL is true, and each vertical carries a current less than $3s$, $f=\pm(c-d)$ or $f=\pm(c+d)$, where f is always less than $3s$. Figs. 6.9 and 6.10 illustrate examples using Figs. 6.8 and 6.7, and Fig. 6.11 illustrates a larger coil for the next case $s=8$.

These coil diagrams show very easily the important conditions:

1) *Each of the numbers* $2, 5, 8, \ldots, 3s-1$ *appears once as a dot in the coil* (and therefore twice in the current graph on two corresponding horizontals).

2) *Each of the numbers* $4, 7, 10, \ldots, 3s-2$ *appears once as a dot in the coil.* (Therefore each appears in the current graph on two corresponding horizontals. The number 1 is used once on the dead-end-arc at vertex x and once as a vertical.)

3) *Each of the numbers* $6, 9, \ldots, 3s-3$ *is a label for two different arches. The number* 3 *is only on one arch* (because the number 3 is used once on the dead-end-arc at vortex y).

4) *The coil begins at dot* 2 *and passes through each of the other dots mentioned in* 1) *and* 2) *except the dot it ends at, which is not dot* 2.

5) *The arch between two dots c and d is labeled with* $\pm(c-d)$ *or* $\pm(c+d)$. (See Fig. 6.12.)

If the conditions 1) to 5) are satisfied, we can use Fig. 6.12 to construct an appropriate current graph. Condition 5) insures KCL. Thus many Cases can be translated into coil problems.

6.5. Orientable Case 9

Consider the current graph of Fig. 6.15 using the group $\mathbb{Z}_{12s+8}$. See also the examples in Figs. 6.13 ($s=0$) and 6.14 ($s=1$). This current graph

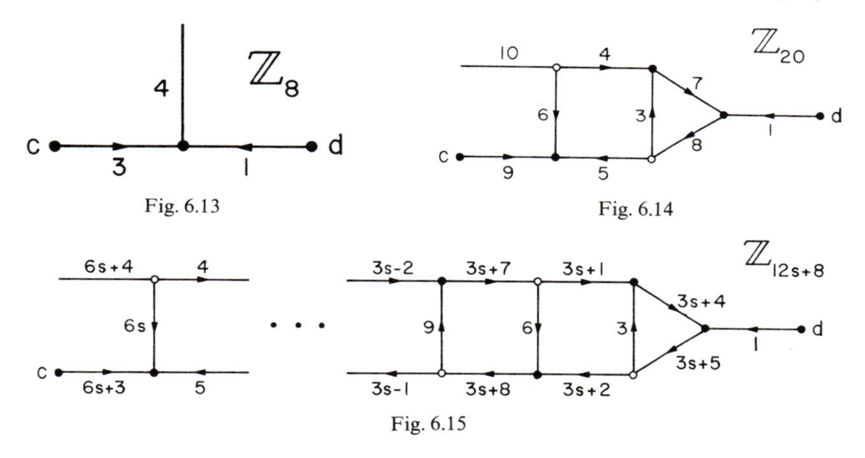

Fig. 6.13

Fig. 6.14

Fig. 6.15

has the properties C1), C2), C4), and C6). The property C3) does not hold because the element 2 does not occur as a current. All the other elements $1, \hat{2}, 3, \ldots, 6s+4$ are used as currents. As usual the notation $\hat{2}$ means that the element 2 is excluded. Therefore row 0 of the produced scheme does not contain the elements 2 and -2. For instance if $s=0$ the produced scheme is (the rows c and d are omitted):

$$
\begin{array}{llllllll}
0. & 5 & c & 3 & 4 & 7 & d & 1 \\
1. & 6 & c & 4 & 5 & 0 & d & 2 \\
2. & 7 & c & 5 & 6 & 1 & d & 3 \\
3. & 0 & c & 6 & 7 & 2 & d & 4 \\
4. & 1 & c & 7 & 0 & 3 & d & 5 \\
5. & 2 & c & 0 & 1 & 4 & d & 6 \\
6. & 3 & c & 1 & 2 & 5 & d & 7 \\
7. & 4 & c & 2 & 3 & 6 & d & 0.
\end{array}
$$

From this scheme we produce a very useful scheme by the following ad hoc operation in row j:

If j is even insert $2+j, x, j-2$ in place of c and omit d.
If j is odd insert $j-2, x, j+2$ in place of d and omit c.

After doing this we obtain the scheme ($s=0$)

$$
\begin{array}{llllllllll}
0. & 5 & 2 & x & 6 & 3 & 4 & 7 & 1 \\
1. & 6 & 4 & 5 & 0 & 7 & x & 3 & 2 \\
2. & 7 & 4 & x & 0 & 5 & 6 & 1 & 3 \\
3. & 0 & 6 & 7 & 2 & 1 & x & 5 & 4 \\
4. & 1 & 6 & x & 2 & 7 & 0 & 3 & 5 \\
5. & 2 & 0 & 1 & 4 & 3 & x & 7 & 6 \\
6. & 3 & 0 & x & 4 & 1 & 2 & 5 & 7 \\
7. & 4 & 2 & 3 & 6 & 5 & x & 1 & 0 \\
x. & (0 & 2 & 4 & 6) & (7 & 5 & 3 & 1).
\end{array}
$$

We will prove that Rule R^* or Δ^* holds in general in the final scheme. Since KCL holds at each vertex of valence 3 we only have to check those places we changed by the ad hoc operation. Row 0 reads before the ad hoc operation:

$$0. \quad \ldots 6s+5, c, 6s+3, \ldots -1, d, 1, \ldots .$$

The ad hoc operation gives us the following part of the scheme:

$$
\begin{array}{rll}
-2. & \ldots 6s+3, 0, x, -4, 6s+1 \ldots & -3, -1 \ldots \\
0. & \ldots 6s+5, 2, x, -2, 6s+3 \ldots & -1, \quad 1 \ldots \\
2. & \ldots 6s+7, 4, x, \quad 0, 6s+5 \ldots & 1, \quad 3 \ldots \\
6s+4. & \ldots 1, 6s+6 \quad . \quad\quad . & \ldots \quad . \quad \ldots \\
6s+6. & \ldots . \quad . \quad . \quad 6s+4, 1 \ldots & . \quad \ldots \\
-1. & \ldots 6s+4, 6s+2 \ldots & -2, -3, x, 1, 0 \ldots \\
1. & \ldots 6s+6, 6s+4 \ldots & 0, -1, x, 3, 2 \ldots \\
3. & \ldots 6s+8, 6s+6 \ldots & 2, \quad 1, x, 5, 4 \ldots \\
6s+3. & \ldots \quad 0, \quad -2 \ldots & . \quad . \quad . \quad . \quad \ldots \\
6s+5. & \ldots \quad 2, \quad\quad 0 \ldots & . \quad . \quad . \quad . \quad \ldots .
\end{array}
$$

Then we can compare row 0 with rows ± 2, ± 1, $6s+5$, and $6s+3$ checking Rule R^* and Δ^*. Also compare row 1 with rows 3, -1, 0, 2, $6s+4$, and $6s+6$. For all the rest Rule Δ^* follows from the fact that the scheme does not change after adding $+2$ to all numbers.

The scheme describes, in dual form, a map with countries 0, 1, 2, ..., $12s+7$, x_0, x_1. The numbered countries are adjacent to each other, while x_0 (x_1) is adjacent to all even (odd) numbered countries. The vertices are all of valence 3. It is easy to compute the genus p of the surface given by the map using Euler's formula.

$$\begin{aligned}
\alpha_2 &= 12s+10\\
2\alpha_1 &= (12+8)(12s+7)+2(12s+8),\\
3\alpha_0 &= 2\alpha_1,\\
6\alpha_0-6\alpha_1+6\alpha_2 &= 12-12p,\\
12p &= 2\alpha_1-6\alpha_2+12,\\
12p &= (12s+8)(12s+9)-6(12s+10)+12,\\
12p &= (12s+6)(12s+5)-6.
\end{aligned}$$

This genus p is exactly one less than the right side of (4.13) for $n=12s+9$. Using one handle we can easily connect the two countries x_0, x_1. This shows that (4.13) holds for all $n\equiv 9 \pmod{12}$.

7. Orientable Cases 11, 2, and 8

7.1. Example for $n=35$

Let the positive integer n be of the form $n=12s+11$. In harmony with the general outline in Section 5.3 we construct a triangular embedding of K_n-K_5 into a closed orientable surface. This means we will find a triangular rotation of K_n-K_5.

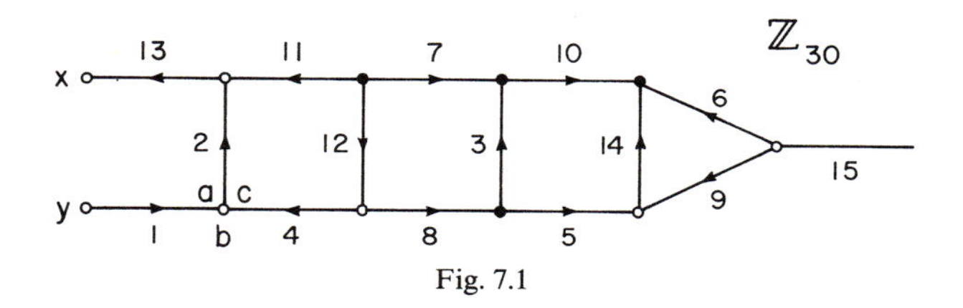

Fig. 7.1

Denote the vertices of $K_{12s+11}-K_5$ by the elements of the group $\mathbb{Z}_{12s+6}$ and the five letters x, y, a, b, c. We will use a current graph. First consider the example for $s=2$ shown in Fig. 7.1. Observe that at the vertex identified by the three letters a, b, c, KCL is not true. But with this exception, the properties C1) to C6) from Sections 2.3 and 6.1 are satisfied. The circuit induced by the rotations of the current graph defines, as in Section 2.3, a row 0. Let us write down the complete log of the circuit using the positive representation of the elements of $\mathbb{Z}_{30}$:

$$15, 6, 16, 25, 22, 18, 11, 13, x, 17, 28, a, 29, y, 1, b, 26, 8,$$
$$3, 23, 12, 4, c, 2, 19, 7, 10, 24, 9, 14, 20, 27, 5, 21.$$

Observe that we put the letters a, b, c at exactly that place where the circuit passes them. For instance, after the arc 4 we put c before continuing with arc 2. We call the vertex where KCL is not true a *vortex* of valence three.

Now regard the log as row 0 of a scheme in which the "numerical" part of row i is obtained by adding $+i$ (mod 30).

A word must be said about the placement of the letters. The positions occupied by x and y are unchanged but the letters a, b, c coming from the vortex of valence 3 are permuted as follows.

0.	…13	x	17	28	a	29	y	1	b	26…4	c	2…
1.	…14	x	18	29	b	0	y	2	c	27…5	a	3…
2.	…15	x	19	0	c	1	y	3	a	28…6	b	4…
3.	…16	x	20	1	a	2	y	4	b	29…7	c	5…
4.	…17	x	21	2	b	3	y	5	c	0…8	a	6…
⋮	⋮	⋮	⋮	⋮	⋮	⋮	⋮	⋮	⋮	⋮	⋮	⋮ .

Reading down through the successive rows under the letters a, b, c there is a cyclic permutation. We have to complete the scheme by saying what the rows x, y, a, b, c are. They are determined by the requirement that rule R^* must be satisfied. This dictates that the rows will be

x.	0	13	26	.	.	.	.	.	.	21	4	17
y.	29	28	27	.	.	.	.	.	.	2	1	0
a.	29	0	28	2	3	1	.	.	.	26	27	25
b.	26	0	1	29	3	4	.	.	.	23	27	28
c.	2	0	4	5	3	7	.	.	.	29	27	1.

Now the scheme is complete. It satisfies rule R^*. Therefore it defines an orientable triangular embedding of $K_{35}-K_5$. Row x or row y is a simple arithmetic sequence with difference 13 or -1 respectively. This comes from the fact that in the current graph the current 13 or -1 flows into the vortex x or y respectively. Both are generators of the group $\mathbb{Z}_{30}$. We already know this type of vortex from Section 2.3 using property C6). We still have to study the rows a, b, c. Observe that the sequence of each third element is an arithmetic sequence independent of where you begin. The difference is always 3.

We used a new construction principle of a vortex of valence 3 to produce the three rows a, b, c. Therefore we have to modify the construction principle C4) slightly and add a new one.

C4*) *At each vertex of valence* 3 *which is not identified by letters, Kirchhoff's Current Law holds.*

C7) *The currents leading into a vortex of valence three denoted by three letters a, b, c (one in each corner) are congruent to one another* (mod 3), *but not congruent to* 0 (mod 3). *The sum of all three currents flowing into the vortex generates the subgroup of all elements divisible by* 3.

In our case where $s=2$ the three currents flowing into the vortex abc are $1, 4$ and -2. Their sum is 3 and $(3, 30)=3$. This means 3 is a generator of the subgroup consisting of the elements 3, 6, 9, ….

We have explained how the construction principle C 7) works if $s=2$ using Fig. 7.1. But we should prove that C 7) works in general: Let us assume that there is a vortex as in Fig. 7.2. We assume that $\alpha\equiv\beta\equiv\gamma\equiv 1$

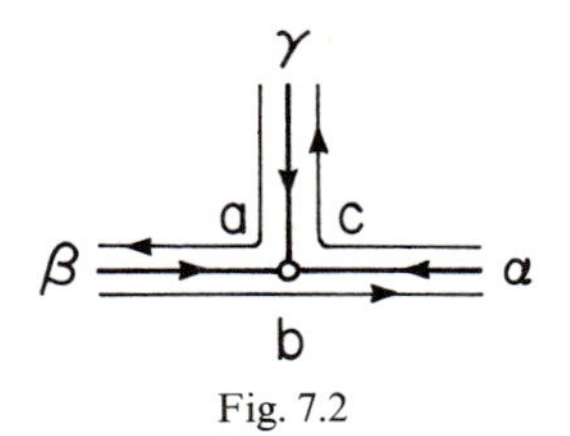

Fig. 7.2

(mod 3). (The other case $\equiv 2$ (mod 3) is entirely analogous. But then in the scheme you have to write c below a, b below c, and a below b. In fact if the currents are $\equiv 1$ (mod 3) you permute the letters in the direction the rotation at the vortex indicates, if they are $\equiv 2$ (mod 3) then you permute in the other direction.) We define $\delta=\alpha+\beta+\gamma$. Fig. 7.2 dictates the following part of row 0 and of the following rows.

$$\begin{array}{rlllll}
0. & \ldots & \gamma,\ a, & -\beta\ldots\beta,\ b, & -\alpha\ldots & \alpha,\ c,\ -\gamma\ldots\\
\gamma. & \ldots & .\ \ . & .\ldots .\ \ . & \ldots. & \delta-\beta,\ a,\ \ 0\ldots\\
\alpha+\gamma. & \ldots & .\ \ . & \ldots\delta,\ a, & \gamma\ldots & .\ \ .\ \ \ldots.\\
\delta. & \ldots\delta+\gamma, & a,\ \delta-\beta & \ldots.\ \ . & \ldots. & .\ \ .\ \ \ldots.\\
\delta+\gamma. & \ldots & .\ \ . & .\ldots .\ \ . & \ldots 2\delta-\beta, & a,\ \ \delta\ldots.
\end{array}$$

Using Rule R^* we get row a:

$$a.\quad \ldots -\beta,\ 0,\ \gamma,\ \delta-\beta,\ \delta,\ \delta+\gamma,\ 2\delta-\beta,\ 2\delta,\ \ldots.$$

A sequence consisting of each third element in the row a is an arithmetic one, for instance, $0, \delta, 2\delta, \ldots$.

Principle C 7) says that δ is a generator of a subgroup with $4s+2$ elements. Therefore row a really contains all elements of the group $\mathbb{Z}_{12s+6}$. The same argument can be used to show that rows b and c will also be "full" cycles.

Referring to Fig. 7.1 there are other rotations which also induce a single circuit, hence other ways to satisfy C 2). There are also other distributions of the elements $1, 2, \ldots, 15$ of $\mathbb{Z}_{30}$ such that construction principles C 1), C 3), C 4*), C 5), C 6), C 7) hold. Each of the above choices gives a scheme for $K_{35}-K_5$ satisfying rule Δ^*. But we really have to find an embedding (certainly not triangular) of the complete graph K_{35}. This is the additional adjacency problem. The embedding of $K_{35}-K_5$

produced by Fig. 7.1 is tailored to fit the requirements imposed by the additional adjacency problem. Before solving it we will present the general Case $n=12s+11$.

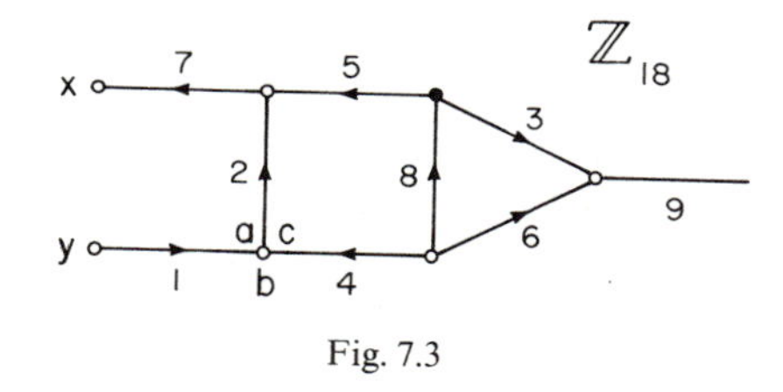

Fig. 7.3

By the way, the diagram in Fig. 7.3 with 2 rungs, as in the example $K_{35}-K_5$, yields a scheme for $K_{23}-K_5$ satisfying rule Δ^*. The trouble is that we cannot complete the additional adjacency problem. Therefore in Section 5.2 we have presented a solution for $n=23$, but it is unfortunately very irregular.

7.2. Orientable Case 11

We will construct a triangular rotation for the graph $K_{12s+11}-K_5$. For this purpose we will use the ladder-like graphs of Figs. 7.4 and 7.5. (Later it will turn out that Fig. 7.4 holds if s is even and Fig. 7.5 if s is odd.) In each case there are $2s$ verticals (rungs) of which three are

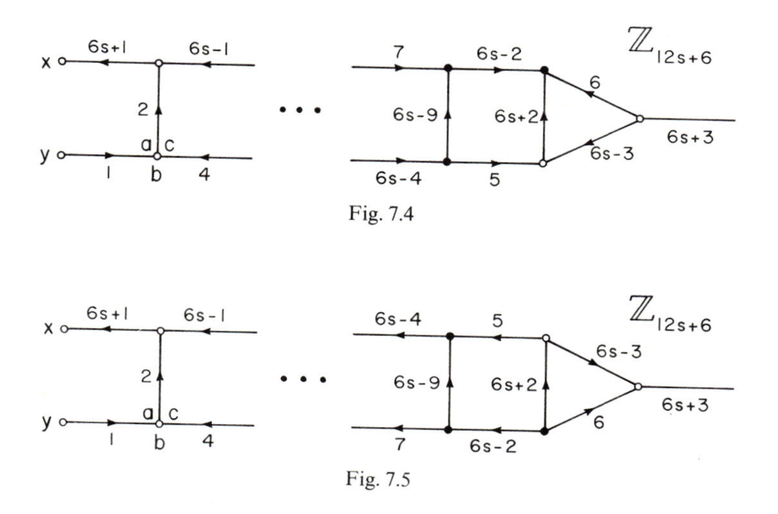

Fig. 7.4

Fig. 7.5

shown. Where rotations are not displayed, place a • on the upper horizontal and ∘ on the lower. These rotations induce a single circuit because there is an even number of verticals.

The graph of Fig. 7.4 (or 7.5) has $6s+3$ arcs. We have to distribute all the unused elements of $\mathbb{Z}_{12s+6}$ from the list

$$1, 2, 3, \ldots, 6s+1, 6s+2, 6s+3 \tag{7.1}$$

on the undisplayed arcs such that KCL is satisfied at each new vertex. This is already the case at all displayed vertices except those at x, y, and the vertex abc. The latter satisfy C5) (1 and $6s+1$ are generators of $\mathbb{Z}_{12s+6}$) and C7).

Consider those elements of the collection (7.1) which are divisible by 3:

$$3, 6, 9, \ldots, 6s-3, 6s, 6s+3. \tag{7.2}$$

Of these the elements 6, $6s-3$, $6s-9$ and $6s+3$ are already displayed, and we have $2s-3$ elements of (7.2) left to place on the diagram. Since there are $2s-3$ undisplayed rungs, this suggests distributing the remaining currents from (7.2) on the undisplayed rungs.

L) *Suppose we have a solution of the distribution problem satisfying* KCL *with currents congruent to* 0 (mod 3) *on the undisplayed rungs in Fig.* 7.4 *or* 7.5.

The horizontals on the ladder must carry elements $\equiv 1$ or $\equiv 2$ (mod 3). Moreover, if t is a horizontal current, then the horizontal current on the *paired* arc directly above or below it must be $6s+3-t$ and point in the same direction as t (both to the right or both to the left). This is because Kirchhoff's Global Current Law (Theorem 2.9) is true and the current at the extreme right of the diagram is $6s+3$.

Hence the currents in (7.1) congruent to 1 or 2 (mod 3) must be considered in pairs

$$3k+1,\ 6s+2-3k \qquad (k=0, 1, \ldots, 2s). \tag{pair k}$$

We call this pair the *pair with number k* or for short the *pair k*.

From the pair 0 the current 1 is used at the extreme left and its partner $6s+2$ on the right-most rung, and from the pair $2s$ both currents are used at the left. The pairs 1, 2, and $2s-1$ also appear in Figs. 7.4 and 7.5, in each case on paired arcs.

Since in the pair 1 the current 4 is on the lower horizontal and points to the left, and since all the currents on the verticals are $\equiv 0$ (mod 3), there are two possibilities for each pair:

If $3k+1$ *is on the lower horizontal then it points to the left. If* $3k+1$ *is on the upper horizontal it points to the right.*

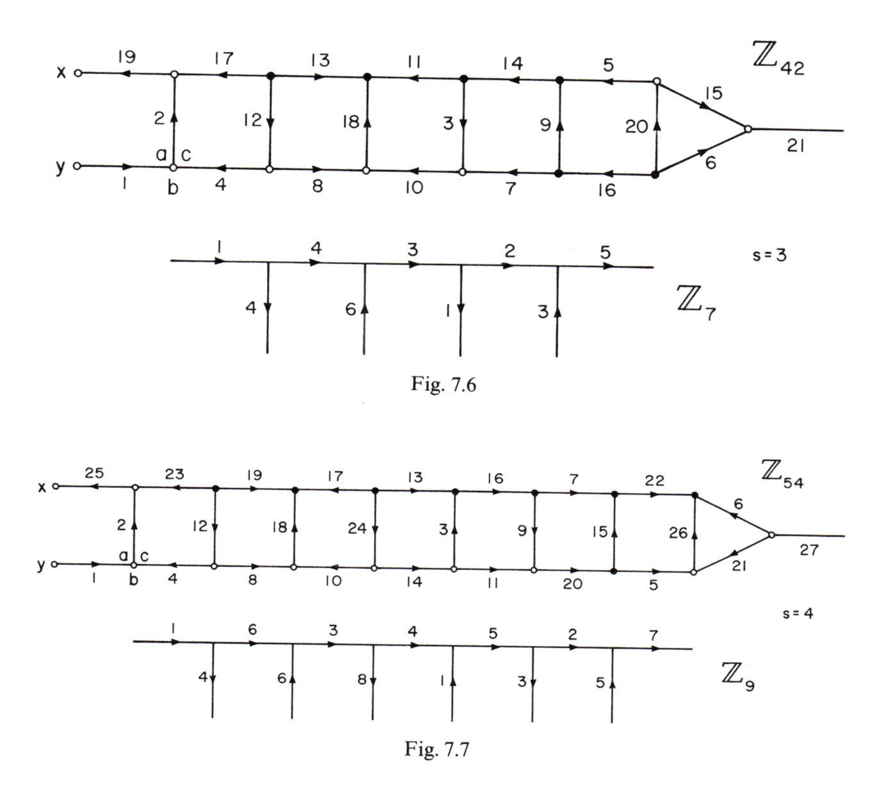

Fig. 7.6

Fig. 7.7

The reader should recall that we are still working under the assumption L). For the next steps refer to the Figs. 7.6 ($s=3$) and 7.7 ($s=4$). For the given ladder-like current graph of Fig. 7.4 (or 7.5) we define an *associated comb* as follows. The comb is drawn right below the ladder, parallel to it, and has $2s-2$ teeth. This is the number of rungs in the ladder having currents $\equiv 0 \pmod 3$. We display currents from the group $\mathbb{Z}_{2s+1}$ on the arcs of the comb using the following Translation Rules.

T1) *If a rung of the ladder carries the current* $3r$ *put the element* r *on the corresponding tooth of the comb and give it the same direction.*

T2) *Let all the horizontals on the comb point to the right.*

T3) *If the current* $3k+1$ *is on a (upper or lower) horizontal of the ladder put the current* k *on the corresponding horizontal of the comb.*

The associated comb is presented in the form

$$
\begin{array}{cccccccccc}
1 & . & . & h & k & . & . & 2 & & 2s-1 \\
 & . & . & . & r & . & . & . & 2s-3 &
\end{array}
\tag{7.3}
$$

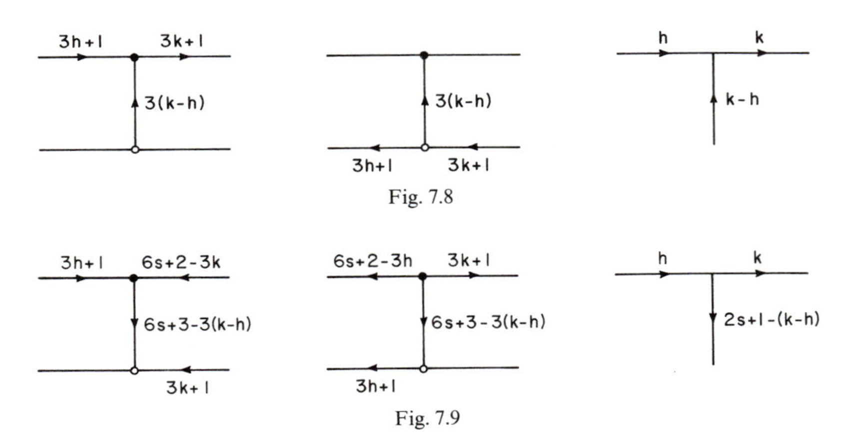

Fig. 7.8

Fig. 7.9

where the first row is the sequence of the horizontals and the second the sequence of the teeth. The following properties of the associated comb are worthy of note, and all but the fourth are easy to check.

A1) *The sequence of horizontals begins with the current* 1 *and ends with* 2, $2s-1$. *The last vertical carries the current* $2s-3$.

A2) *All the arrows on the horizontals point to the right. The currents on the horizontals (these are pair numbers) are distinct and exhaust the collection* $\hat{0}, 1, \ldots, 2s-1, \widehat{2s}$. *As usual the notation* $\hat{a}$ *means that the element* a *is to be excluded.*

A3) *The currents on the teeth are distinct and exhaust the collection* $1, \hat{2}, 3, \ldots, 2s-2, \widehat{2s-1}, 2s$.

A4) *At each vertex of valence* 3 *in the comb,* KCL *holds in the group* $\mathbb{Z}_{2s+1}$.

In order to prove A4) we assume that in the associated comb the two elements h, k are two successive currents and $h<k$. According to the translation rule T3) it follows that $3h+1$ and $3k+1$ are found on the corresponding upper and/or lower horizontals on the ladder. That means that one of the four partial pictures of the ladder shown in Fig. 7.8 (left and middle part) and in Fig. 7.9 (left and middle part) represent the situation.

The currents on the verticals in the ladder are dictated by KCL. Consequently the right-hand partial picture of Fig. 7.8 or Fig. 7.9 appears in the associated comb. In both cases KCL holds according to the group $\mathbb{Z}_{2s+1}$.

In the case $k<h$ the same argument can be used. Just write $h-k$ instead of $k-h$ and reverse the arrows on all the verticals in Figs. 7.8 and 7.9.

Up to this point we have assumed the statement L). Let us summarize what we have shown:

If we have a solution of the distribution problem for the ladder-like graph of Fig. 7.4 or 7.5 satisfying KCL in $\mathbb{Z}_{12s+6}$ with currents $\equiv 0 \pmod 3$ on the undisplayed rungs, then there exists a comb with the properties A 1), A 2), A 3), and A 4).

The converse statement is also true and is even more important for our purpose.

If there exists a comb having the properties A 1), A 2), A 3), and A 4) then it is possible to distribute all the unused elements of $\mathbb{Z}_{12s+6}$ on the undisplayed arcs of Fig 7.4 or 7.5 such that KCL is satisfied at each new vertex.

Proof. Suppose there is a given comb having the properties A 1), A 2), A 3), and A 4). We have to fill out the ladder in Fig. 7.4 (or 7.5). Using the translation rule T 1) in the reverse direction it is easy to find the currents on the rungs. Just multiply the currents of a tooth of the comb by three to get the current of the corresponding rung of the ladder. Give the rungs the same orientation as the teeth.

But how do we find the currents on the horizontals? The rule T 3) does not determine them uniquely. But with the help of KCL it will. We have to begin filling in the horizontals of the ladder at the *left* side.

Let $1, c_2, c_3, \ldots, c_i, h, k, \ldots$ be part of the sequence of currents on the horizontals of the given comb. Moreover let us assume we have already filled in the horizontal currents up to the pair h: $3h+1, 6s+2-3h$. So $3h+1$ is on the upper or lower horizontal. The property A 4) says that if $h<k$ then on the intervening tooth we find $k-h$ or $2s+1-(k-h)$. Therefore one of the four possibilities of Figs. 7.8 and 7.9 will hold and determines how to fill in the next pair k of the ladder so that KCL holds.

The argument is analogous if $k<h$. Proceeding in this fashion we come up to the right where the pair 2 is already displayed. We cannot yet see whether we shall end as in Fig. 7.4 or as in Fig. 7.5. In any event the only question left is how to construct a comb having properties A 1), A 2), A 3), and A 4). In Figs. 7.6 and 7.7 the solutions for $s=3$ and 4 are presented. The next two examples $s=5$ and 6 are as follows.

$$\begin{array}{ccccccccccccccccccc} 1 & & 8 & & 3 & & 6 & & 5 & & 4 & & 7 & & 2 & & 9 \\ & 4 & & 6 & & 8 & & 10 & & 1 & & 3 & & 5 & & 7 & \end{array} \qquad (\mathbb{Z}_{11})$$

$$\begin{array}{ccccccccccccccccccccc} 1 & & 10 & & 3 & & 8 & & 5 & & 6 & & 7 & & 4 & & 9 & & 2 & & 11 \\ & 4 & & 6 & & 8 & & 10 & & 12 & & 1 & & 3 & & 5 & & 7 & & 9 & \end{array} \qquad (\mathbb{Z}_{13})$$

Recall that the arrows on the horizontals all point to the right. The arrows on the verticals are determined by KCL. As a matter of fact they alternate in direction. Therefore the present combs are really arithmetic combs in the sense of Section 6.2. (Referring to the last comb above, notice that $12+2\equiv 1 \pmod{13}$.)

The generalization for any given s is obvious. The sequence of the teeth (verticals) is

$$4, 6, 8, \ldots, 2s, 1, 3, 5, \ldots, 2s-3$$

while the sequence of horizontal currents is given by dovetailing the sequence

$$1, 3, 5, \ldots, 2s-1$$

with the sequence

$$2s-2, 2s-4, 2s-6, \ldots, 2.$$

The even currents of the teeth are all of the type in Fig. 7.9. That means they manufacture a changing of the currents which are $\equiv 1 \pmod 3$ in the ladder from lower to upper horizontal or vice versa. Since there are $s-1$ teeth with even currents this means that if s is even we end as in Fig. 7.4 and if s is odd as in Fig. 7.5.

Remark. We can easily get lots of other solutions. For example from the solution for $s=5$ we obtain another one by interchanging the numbers 8 and 3 in the sequence of the teeth. There are also completely different solutions which are rather peculiar. An example for $s=5$:

$$\begin{matrix} 1 & & 5 & & 4 & & 7 & & 6 & & 3 & & 8 & & 2 & & 9 \\ & 4 & & 1 & & 3 & & 10 & & 8 & & 5 & & 6 & & 7 & \end{matrix} \qquad (\mathbb{Z}_{11})$$

7.3. The Additional Adjacency Problem

The previous section gives us a scheme for $K_{12s+11}-K_5$ satisfying Rule Δ^* if $s\geqq 2$. This means that there exists a triangular embedding of $K_{12s+11}-K_5$ into an orientable surface. There are 10 missing adjacencies needed to complete K_{12s+11}. Theorem 5.7 suggests that we should try to manufacture these 10 missing arcs with the help of $\{10/6\}=2$ additional handles on the surface.

First consider the dual map of the triangular embedding of $K_{12s+11}-K_5$. This map has $12s+11$ countries identified by $0, 1, \ldots, 12s+5, x, y, a, b, c$. Each vertex is incident with three countries and three arcs.

Any two countries are adjacent unless both countries are in the exceptional set of the five countries x, y, a, b, c.

The information required for the solution of the additional adjacency problem is obtained from the displayed parts of Fig. 7.4 (if s

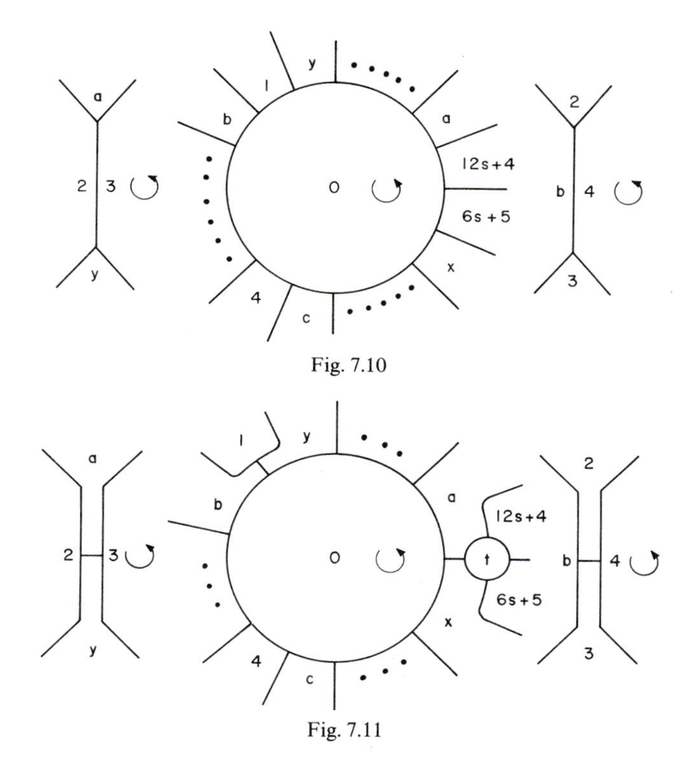

Fig. 7.10

Fig. 7.11

is even) or 7.5 (if s is odd). The left side of the figures gives us part of row 0 as recorded below and, in addition, useful parts of other rows.

0.	x	$6s+5$	$12s+4$	a	$12s+5$	y	1	b	$12s+2$	… 4 c …
1.	.	.	$12s+5$	b	0	y	2	c	.	…
2.	.	.	0	c	1	y	3	a	.	…
3.	.	.	1	a	2	y	4	b	.	…
4.	.	.	2	b	3	y	5	c	.	…
.	.	.	.	.	.	.	.	.	.	…
.	.	.	.	.	.	.	.	.	.	…
$12s+4$.	.	.	.	.	.	y	$12s+5$	c	$12s$	…
$12s+5$.	.	.	.	.	.	y	0	a	$12s+1$	….

This provides a partial picture (Fig. 7.10) of the map around the countries 0, 3, and 4.

First Modification. We modify the map as illustrated in Fig. 7.11 (subdivisions and compositions of dimension 1 and 2). The new country

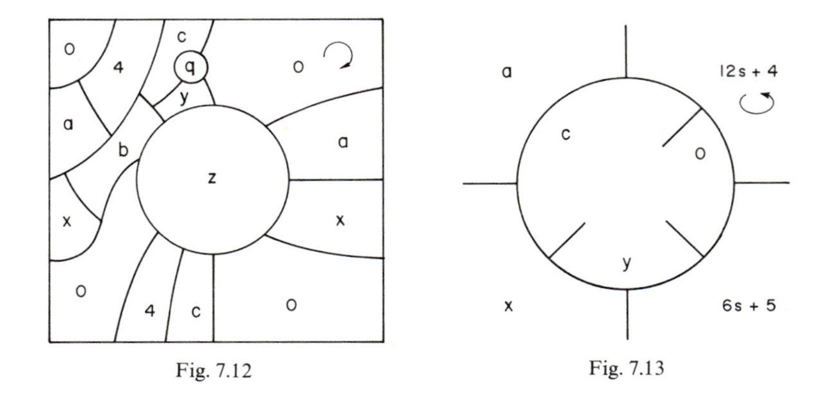

Fig. 7.12

Fig. 7.13

identified by t we will use later. Notice that we have gained the three adjacencies

(7.4) $$(a, y),\ (b, y),\ (a, x)$$

but lost the four adjacencies

(7.5) $$(0, 1),\ (0, 12s+4),\ (0, 6s+5),\ (b, 4).$$

The adjacency (2, 3) merely changed its position in the map.

First Handle. We consider the map of Fig. 7.12 on a torus. Notice that 0 appears to be cut in four parts but is, in fact, a single country.

Now excise the country 0 in Fig. 7.11 and the country z in Fig. 7.12. Then identify the two boundaries of the two resulting partial polyhedra in the obvious way. After this there is a new country named 0, that is adjacent to the same countries as the old country 0 in Fig. 7.11.

Using the first handle we gained the six adjacencies

(7.6) $$(a, b), (b, c), (b, x), (c, y), (b, 4), (a, 4).$$

The last adjacency $(a, 4)$ is *extra* in the sense that it already exists elsewhere on the map.

Second Handle. Next excise the country t in Fig. 7.11 and the country q in Fig. 7.12. Identify the boundaries of the resulting two partial polyhedra as illustrated in Fig. 7.13. Notice that this creates a closed surface which is orientable. We gained seven adjacencies

(7.7) $$\begin{array}{l} (a, c), \quad (c, x), \quad (x, y), \\ (0, 12s+4), (0, 6s+5), (y, 6s+5), (c, 12s+4). \end{array}$$

The last two are extra.

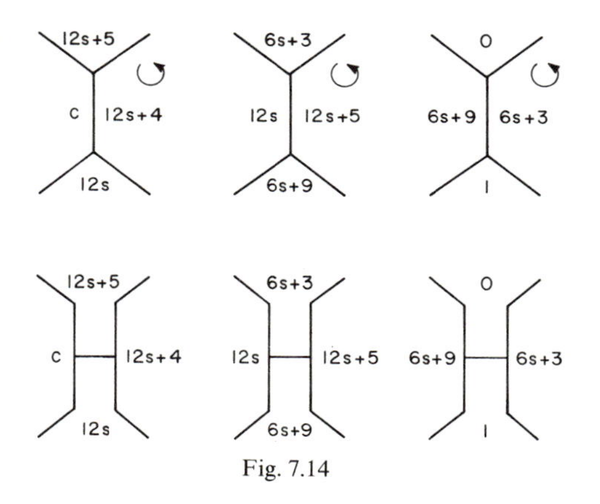

Fig. 7.14

From (7.4), (7.6) and (7.7) we see that all the countries identified by letters a, b, c, x, y are now adjacent to each other. In doing this we have lost the adjacencies listed in (7.5). But in (7.6) we regained the last adjacency $(b, 4)$, and in (7.7), the last adjacencies $(0, 12s+4)$ and $(0, 6s+5)$. Hence only $(0, 1)$ remains missing.

Final Modification. If s is even then the right side of Fig. 7.4 gives the following part of row 0.

$$0. \quad \ldots 6s+3 \quad 6 \quad 6s+4 \quad 12s+1 \quad 6s+10\ldots.$$

Using Fig. 7.5 if s is odd gives:

$$0. \quad \ldots 6s+10 \quad 12s+1 \quad 6s+4 \quad 6 \quad 6s+3\ldots.$$

One is obtained from the other by a complete reversal in the order of the five numbers involved. If s is even we have

$$\begin{array}{llllll} 6s+3. & \ldots 0 & 6s+9 & 1 & \ldots & \\ 12s+5. & \ldots & \ldots & 6s+3 & 12s & 6s+9\ldots. \end{array}$$

Using the above and row $12s+4$ from the earlier displayed portion of the scheme we obtain the local picture in the upper section of Fig. 7.14 if s is even. If s is odd reverse the orientations displayed in the center and at the right of Fig. 7.14.

In using the second handle we gained the extra adjacency $(c, 12s+4)$ (see (7.7)). Fig. 7.14 (upper left) shows the original place of the adjacency $(c, 12s+4)$. Since we now have it elsewhere we no longer need it here. So we modify as shown in Fig. 7.14 (lower left) and gain $(12s, 12s+5)$. Now we no longer need the original place of the adjacency $(12s, 12s+5)$ shown in Fig. 7.14 (upper central). Modify it as in the lower central part of the figure. We gain $(6s+3, 6s+9)$. Therefore we can modify the right partial picture as shown in the lower right part of Fig. 7.14. So we recover the remaining lost adjacency (0, 1).

7.4. Orientable Case 2

The ladder-like current graph we will use is very easy to describe because it is almost the same as in the Case 11. But the additional adjacency problem will be rather adventurous.

In order to be able to use the notation of Case 11 let n be given in the form $n=12s+14$.

Consider the ladder-like current graph of Fig. 7.15 if s is even and of Fig. 7.16 if s is odd. In both cases the ladder has $2s$ rungs. The currents are elements of the group $\mathbb{Z}_{12s+6}$.

Fig. 7.17 illustrates the example $s=2$. Compare Fig. 7.15 with Fig. 7.4, and notice that the six elements 1, $6s+2$, 2, $6s+1$, 6, $6s-3$ have changed places. All other elements are displayed alike in both figures. We even assume that the undisplayed rungs and horizontals of Fig. 7.15 carry exactly the same currents as in Fig. 7.4 (s is even). In Section 7.2 we explained how to complete the current graph of Fig. 7.4. Is s is odd consider Fig. 7.16 and take the currents of all the undisplayed rungs and horizontals from Fig. 7.5.

Where rotations in Figs. 7.15 and 7.16 are not specified place a clockwise rotation (filled in circles) on the upper horizontal and a counterclockwise rotation (empty circles) on the lower.

Then the current graph has the properties:

C1) Each vertex has valence 3 or 1.

C2) The given rotation induces one single circuit.

C3) Each element of the group except 0 appears exactly once in the circuit.

C4*) At each vertex of valence 3, which is not identified by letters, Kirchhoff's Current Law is true.

C5*) The element $6s+1$ on the end arc at vertex x generates the group $\mathbb{Z}_{12s+6}$. But the element 2 on the end arc at vertex y generates the subgroup of all even elements of $\mathbb{Z}_{12s+6}$.

At the two vertices abc and uvw the property C7) of Section 7.1 holds: The currents flowing into vertex uvw are 2, $6s-1$, and $6s+2$. All three are $\equiv 2 \pmod 3$. The sum $2+6s-1+6s+2=12s+3=-3$ generates the subgroup of all multiples of 3. The currents flowing into vortex abc are 4, $6s+1$, and $-(6s+2)$. All three are $\equiv 1 \pmod 3$. Their

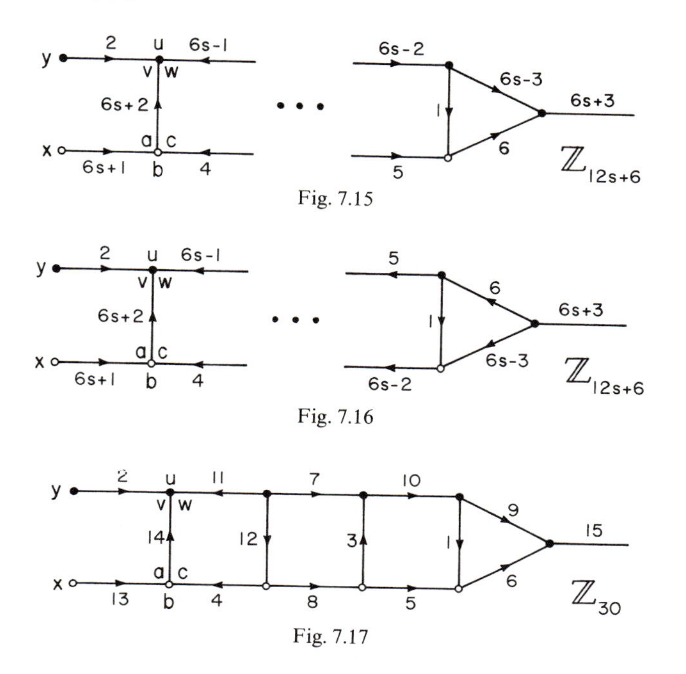

Fig. 7.15

Fig. 7.16

Fig. 7.17

sum is 3 and generates the same subgroup of index 3. The letters will be permuted as explained in Section 7.1.

The circuit induced by the rotation of the current graph in Fig. 7.15 or 7.16 gives rise to a row 0 for a scheme.

Next we construct the rows 1, 2, etc. by the additive rule, but x and y are unchanged, and the letters a, b, c and u, v, w are permuted as

$$
\begin{array}{lll}
0. & \ldots a \ldots & \ldots u \ldots \\
1. & \ldots b \ldots & \ldots v \ldots \\
2. & \ldots c \ldots & \ldots w \ldots \\
3. & \ldots a \ldots & \ldots u \ldots
\end{array}
$$

etc. Then the rows a, b, c, u, v, w, x, y are determined by Rule Δ^* according to principle C7) from Section 7.1 and C5) from Section 2.3.

Since the current 2 on the dead-end-arc at vertex y is not a generator we cannot expect to get a normal row for y. In fact, when we try to construct row y using Rule Δ^* we get two cycles, namely,

$$y.\quad (12s+4 \ldots 4\ 2\ 0)\quad (12s+5 \ldots 5\ 3\ 1).$$

As we did in Section 5.2 in the scheme 5.7 ($n=8$ or 9) we replace y by y_0 in the rows 0, 2, ... and by y_1 in the rows 1, 3, ... We obtain a triangular rotation for a graph G_{12s+15} which can be described as follows: there are $12s+15$ vertices denoted by $0, 1, 2, \ldots, 12s+6, a, b, c, u, v, w, x, y_0, y_1$.

Each pair of vertices identified by numbers is joined by an arc; the vertices a, b, c, u, v, w and x are joined to all numbered vertices; the vertex y_0 is joined to all the even numbered vertices, y_1 to all the odd ones. Notice that there are no arcs connecting any pairs of vertices identified by letters.

Remark. This graph G_{12s+15} appears to be very unlike the complete graph K_{12s+14}. But if in the graph y_0 and y_1 are identified so as to obtain a single vertex y, then we obtain the graph $K_{12s+14}-K_8$.

The scheme obtained provides a triangular embedding of the graph G_{12s+15} into an orientable surface S. We can consider the dual map M on S which has all vertices of valence three. Since M has

$$\alpha_2 = 12s+15 \text{ countries and}$$

$$\alpha_1 = \binom{12s+6}{2} + 8(12s+6) = 72s^2 + 162s + 63 \text{ arcs}$$

the genus p of S can be determined using Theorem 4.4:

$$\alpha_1 = 3\alpha_0 - 3E(S) = 3\alpha_0 - 6 + 6p$$

$$p = 12s^2 + 21s + 4.$$

So we have found a sophisticated map M with $12s+15$ countries on the orientable surface of genus $12s^2+21s+4$.

But we really wish to have a map with $n=12s+14$ mutually adjacent countries on the orientable surface of genus

$$\left\{\frac{(n-3)(n-4)}{12}\right\} = \left\{\frac{144s^2+252s+110}{12}\right\} = 12s^2+21s+10.$$

Notice that the difference between the two genera is six. That means for the map M the additional adjacency problem is the following: Use 6 additional handles in order to manufacture the 28 missing adjacencies (all pairs of the eight countries a, b, c, u, v, w, x, and y_0 are non-adjacent) and one more adjacency between y_0 and y_1. Then you can cancel the boundary between y_0 and y_1 and call the new country y. The result will be a map with $12s+14$ mutually adjacent countries. We will do all this in the next section.

7.5. Additional Adjacency Problem

Consider the following part of row 0, which can easily be obtained from Fig. 7.15 or 7.16.

$$\begin{array}{lcccccccccccc} 0. & \ldots 6s-1 & w & 6s+4 & a & 6s+5 & x & 6s+1 & b & 12s+2\ldots \\ & \ldots \quad 4 & c & 6s+2 & v & 12s+4 & y_0 & 2 & u & 6s+7\ldots. \end{array}$$

The following part of the scheme is also important.

$$\begin{array}{llll} 0. & \ldots 6s+2 & v & 12s+4\ldots \\ 1. & \ldots 6s+3 & w & 12s+5\ldots \\ 2. & \ldots 6s+4 & u & 0 \quad \ldots. \end{array}$$

This gives a partial picture of the map M around the countries 0 and 2 as illustrated in Fig. 7.18.

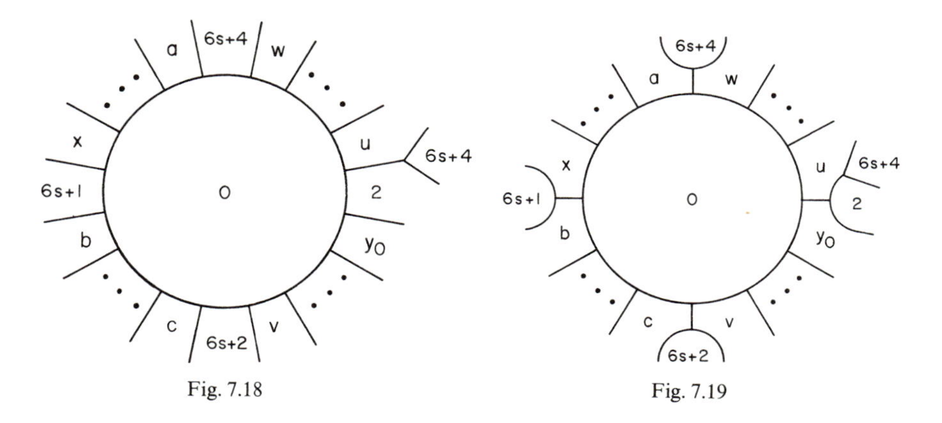

Fig. 7.18

Fig. 7.19

First Modification. We modify the map as illustrated in Fig. 7.19. We gain the adjacencies

$$(a, w), (b, x), (c, v), (u, y_0)$$

but we lose the adjacencies

$$(0, 2), (0, 6s+1), (0, 6s+2), (0, 6s+4). \tag{7.8}$$

First Handle. Consider the map on a torus shown in Fig. 7.20 and ignore the two dotted circles. Excise country 0 from Fig. 7.19 and country z from Fig. 7.20. Identify the boundaries of the two resulting partial polyhedra in the obvious way. After this there is a new country, named 0, that is adjacent to the same countries as the old country 0 in Fig. 7.19.

We have gained the adjacencies

$$(a, c), (b, y_0), (x, u), (v, w).$$

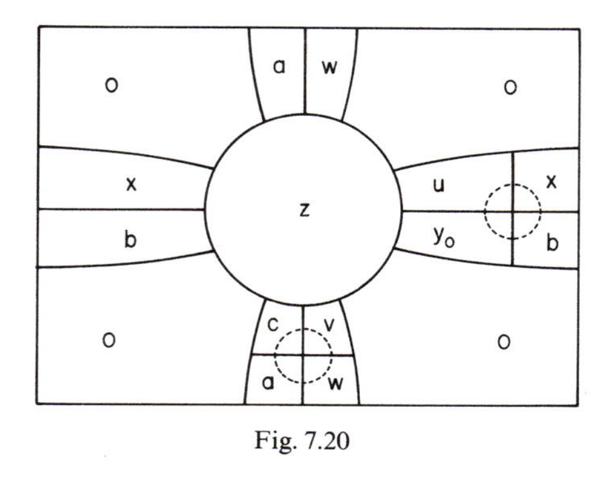

Fig. 7.20

Second Handle. From Fig. 7.20 excise the interiors of the two dotted circles and identify the boundaries as in Fig. 7.21. (Now ignore the dotted circles in Fig. 7.21.) This identification is made so that the new surface is orientable. The same will be done in identifications to follow. We gain the adjacencies

$$(a, b),\ (b, w),\ (w, y_0),\ (y_0, v),\ (v, u),\ (u, c),\ (c, x),\ (x, a).$$

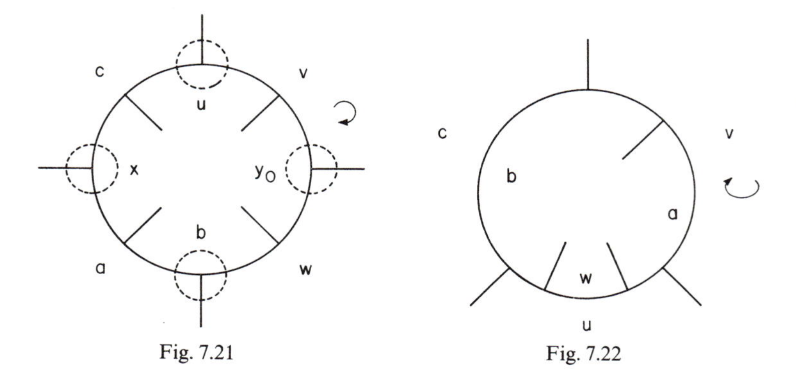

Fig. 7.21

Fig. 7.22

Third Handle. Excise the interiors of the upper and lower dotted circles in Fig. 7.21. Identify the boundaries as in Fig. 7.22. We gain the adjacencies

$$(a, u),\ (u, b),\ (b, v),\ (v, a),\ (b, c),\ (u, w).$$

Fourth Handle. Now excise the interiors of the dotted circles to the left and right of Fig. 7.21. Identify the two boundaries as in Fig. 7.23. We gain the six adjacencies

$$(c, y_0),\ (y_0, x),\ (x, w),\ (w, c),\ (a, y_0),\ (v, x).$$

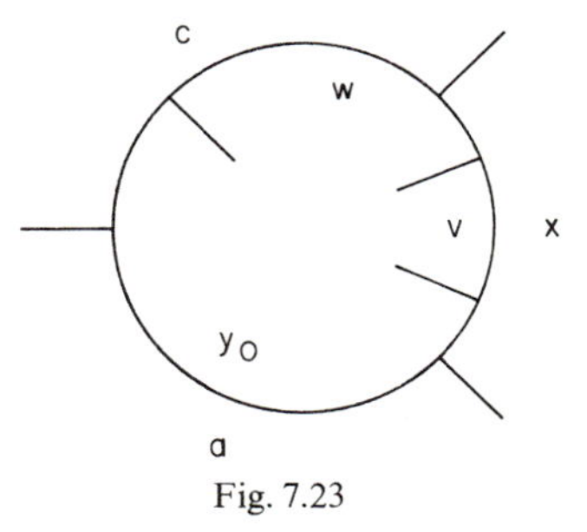

Fig. 7.23

What we gained by using only four handles is certainly surprising. All 28 adjacencies between the countries a, b, c, u, v, w, x, y_0 are now obtained. On the other hand, we have lost the four adjacencies of (7.8).

Fifth Handle. From Fig. 7.19 and from the fact that country u is adjacent somewhere to $6s+2$ we obtain a partial picture shown in Fig. 7.24. Consider Fig. 7.25 as a map on a torus. Excise country u from Fig. 7.24 and country z from Fig. 7.25 and identify in the obvious way. Again the old country u is replaced by a new country u. We gained the adjacencies

$$(0, 2), (0, 6s+2), (0, 6s+4).$$

Hence we have recovered three of the adjacencies (7.8) lost in the first modification.

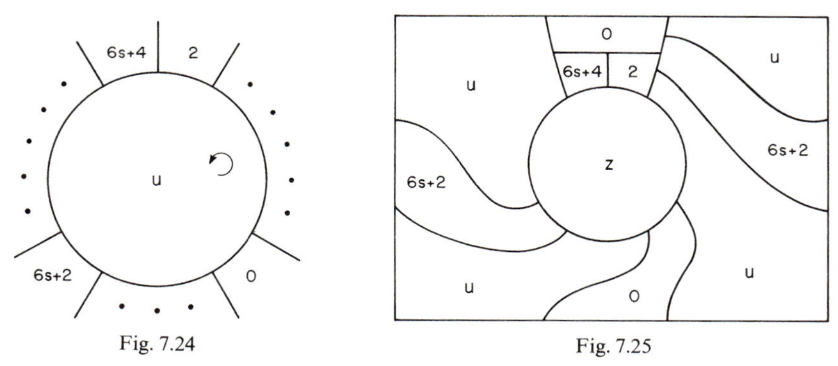

Fig. 7.24

Fig. 7.25

Sixth Handle. Now we need only recover the last missing adjacency $(0, 6s+1)$ and manufacture an extra adjacency between y_0 and y_1. The country y_1 is adjacent to all the odd-numbered countries. Therefore y_1 and $6s+1$ have a common boundary. Likewise y_0 and 0 have a common boundary. It is now easy by the addition of one handle to obtain the two adjacencies $(0, 6s+1)$ and (y_0, y_1) simultaneously.

7.6. Orientable Case 8

It is now easy to explain the solution in Case 8 because we can use all the ideas of the two previous Cases 11 and 2. In fact we will again use the same group $\mathbb{Z}_{12s+6}$.

Consider the ladder-like current graph of Fig. 7.26 (upper part) where the total number of rungs is $2s$. (Fig. 7.27 illustrates an example for $s=2$.) Where rotations are not displayed place ● on the upper horizontal and ○ on the lower. This rotation of the graph produces one single circuit.

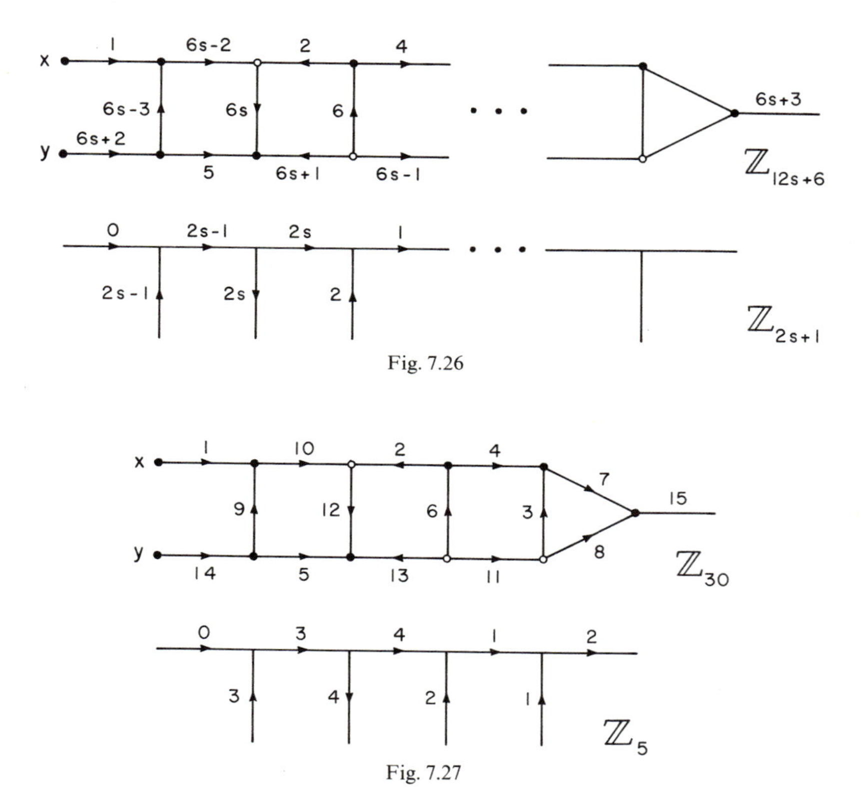

Fig. 7.26

Fig. 7.27

The current graph in Fig. 7.26 has $6s+3$ arcs. We intend to distribute the elements

$$1, 2, 3, \ldots, 6s+3$$

as currents on these arcs such that KCL is satisfied at each vertex of valence three. Some of the elements are already displayed. As in the two previous cases we will put all the multiples of 3 on the verticals. That means the currents of any two corresponding horizontals are of the form

$$3k+1,\ 6s+2-3k \qquad (k=0, 1, 2, \ldots, 2s).$$

Remember this follows from the global KCL (Theorem 2.9). This pair we have called the pair k. Consider the two arcs on the triangle of the ladder on the right also as "horizontals".

Suppose we have a solution of the described distribution problem for the current graph (Fig. 7.26, upper part). Then we can construct an associated comb as follows (Fig. 7.26, lower part).

T0) The comb has $2s$ rungs.

T1) If $3r$ is the current of a vertical in the current graph put the element r on the corresponding rung of the comb and choose the same orientation as in the vertical.

T2) On the horizontals of the comb let all the arrows point to the right.

T3) If the current $3k+1$ is on an (upper or lower) horizontal of the ladder put the current k on the corresponding horizontal of the comb.

In Section 7.2 we proved that the constructed comb has the properties A1 to A4:

A1) The left side of the comb begins with

$$\begin{array}{ccccccccc} 0 & & 2s-1 & & 2s & & 1 & \ldots \\ & 2s-1 & & 2s & & 2 & & \ldots. \end{array}$$

A2) All the arrows on the horizontals point to the right. The currents on the horizontals are distinct and exhaust the collection $0, 1, 2, \ldots, 2s$.

A3) The currents on the teeth of the comb are distinct and exhaust the collection $1, 2, \ldots, 2s$.

A4) At each vertex of valence 3 of the comb KCL holds in the group $\mathbb{Z}_{2s+1}$.

In Section 7.2 we have seen that the reverse construction can be made: *If we can find a comb with $2s$ teeth and the properties* A1), A2), A3), *and* A4), *then the distribution problem for the ladder-like graph of Fig.* 7.26 *can be solved.*

So the problem which is left is how to find such a comb for each value of s. Fig. 7.27 illustrates an example for $s=2$. For $s=4$, 6, and 8 see the examples:

$$\begin{matrix} 0 & 7 & 8 & 1 & 5 & 2 & 6 & 3 & 4 \\ 7 & 8 & 2 & 5 & 6 & 4 & 3 & 1 & \end{matrix} \qquad \binom{s=4}{\mathbb{Z}_9}$$

$$\begin{matrix} 0 & 11 & 12 & 1 & 9 & 2 & 10 & 3 & 7 & 4 & 8 & 5 & 6 \\ 11 & 12 & 2 & 6 & 5 & 8 & 7 & 9 & 10 & 4 & 3 & 1 & \end{matrix} \qquad \binom{s=6}{\mathbb{Z}_{13}}$$

$$\begin{matrix} 0 & 15 & 16 & 1 & 13 & 2 & 14 & 3 & 11 & 4 \\ 15 & 16 & 2 & 5 & 6 & 12 & 11 & 9 & 10 & \\ & & & 12 & 5 & 9 & 6 & 10 & 7 & 8 \\ & & & 8 & 7 & 13 & 14 & 4 & 3 & 1\,. \end{matrix} \qquad \binom{s=8}{\mathbb{Z}_{17}}$$

In these three examples we did not say what the arrows on the arcs are. But these are easily determined by A2) and KCL. The generalization of these examples for $s=1$, 4, 6, and 8 to any given *even* number s is not difficult to describe.

$$\begin{matrix} 0 & 2s-1 & 2s & 1 & 2s-3 & 2 & 2s-2 \\ 2s-1 & 2s & 2 & 5 & 6 & 2s-4 & \\ & & 3 & 2s-5 & 4 & 2s-4\ldots & \\ & 2s-5 & 9 & 10 & 2s-9 & \ldots & \\ & \ldots s-2 & s+2 & s-1 & s & & \\ & \ldots & 4 & 3 & 1 & & \end{matrix} \qquad (\mathbb{Z}_{2s+1})$$

But we prefer a more geometric way to explain the general pattern of the solution. First some remarks:

If we interchange for instance the numbers 4 and 5 in the lower row of the solution for $s=4$ we get another solution for the same problem.

In all the given solutions the number of downward pointing teeth equals the number of teeth pointing up.

Instead of the problem of finding a comb with the properties A1), A2), A3), and A4), we look at another similar question. It is really a coil diagram in a cyclic form.

Consider $2s+1$ points on the circumference of a circle which form the vertices of a regular $(2s+1)$-gon. Denote the vertices by $0, 1, 2, \ldots, 2s$. The *length* of a chord connecting two vertices k and h is defined to be

$$\min(|k-h|, 2s+1-|k-h|).$$

Hence the greatest possible length of a chord is s. Now we consider the following

Chord Problem. Draw a continuous sequence of chords with the properties:

Ch 1) *The sequence begins at point* 0.

Ch 2) *The sequence passes through every vertex exactly once.*

Ch 3) *Each possible length appears exactly twice in the sequence of chords.*

This Chord Problem is very easy to solve. A solution is given by Fig. 7.28 for $s=6$. The lengths of the chords used are 1, 2, 3, 4, 5, 6, 6, 5, 4, 3, 2, 1 in this order. The generalization of this solution for an arbitrary number s is obvious. We call a solution (or a part of it) a *simple zigzag* if the length of the chords increases or decreases by one and the angle of two successive chords is always acute.

Since this chord problem was too easy we make it a little more difficult.

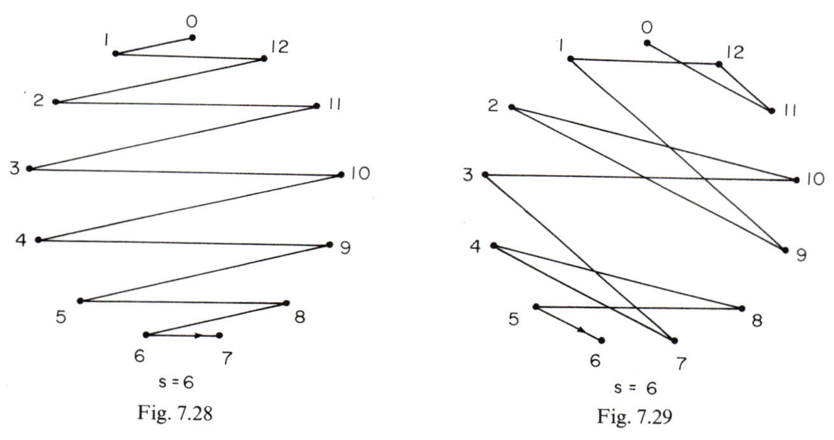

Fig. 7.28 Fig. 7.29

Aggravated Chord Problem. Find a sequence of chords with properties Ch 2), Ch 3) *and property*

Ch 1*) *The sequence begins with*

$$0, 2s-1, 2s, 1, \ldots .$$

Fig. 7.29 shows a solution for $s=6$. Before we discuss a general solution we will prove *that each solution of the Aggravated Chord Problem provides a solution of the comb problem.*

Consider the given sequence of chords. There are two chords of length k where $k=1, 2, \ldots, s$. Label one of the two chords k, the other $2s+1-k$.

Now construct a comb with $2s$ teeth. The currents on the horizontals take the numbers of the vertices in the given sequence and the arrows on the horizontals all point to the right. This takes care of property A 2). The currents on the teeth take the labels of the chords in the same order as they appear in the given sequence of chords. Therefore A 3) is true.

If h and k are the currents of two successive horizontals and $h<k$ then the current of the interposed tooth is $k-h$ or $2s+1-k+h$. In the first case let the arrow of the tooth point up. In the second case let the arrow point down. If $k<h$ then choose the opposite orientation of the tooth. In any case KCL is true and therefore A 4) holds.

The first three chords have the length 2, 1, 2. We are allowed to label them by $2s-1$, $2s$, 2. Therefore the comb can be constructed such that property A 1) is also satisfied.

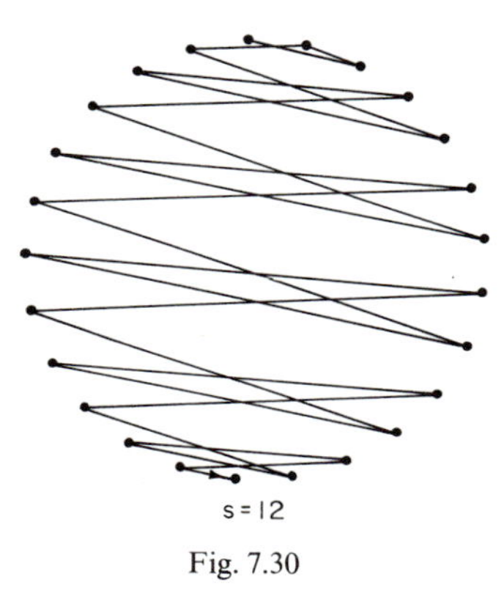

Fig. 7.30

Fig. 7.30 illustrates how the general solution for the aggravated chord problem for even s can be made. The order of the length of the chords is

$$2, 1, 2;\ 5, 6, 5, 6;\ 9, 10, 9, 10;\ \ldots;\ 4, 3, 4, 3;\ 1.$$

Such a solution we like to call a "fox trot" solution because after four steps the pattern is repeated.

Now we go to the case where s is an odd number. There is no solution if $s=1$ or 3. For $s=5$ the comb

$$\begin{matrix} 0 & & 9 & & 10 & & 1 & & 6 & & 2 & & 5 & & 4 & & 7 & & 3 & & 8 \\ & 9 & & 10 & & 2 & & 5 & & 4 & & 3 & & 1 & & 8 & & 7 & & 6 & \end{matrix} \qquad (\mathbb{Z}_{11})$$

solves the problem. For some of the next cases see Fig. 7.31. It is easy to see how the general pattern works. The solution is composed of three simple zigzags. The one in the middle always has five chords.

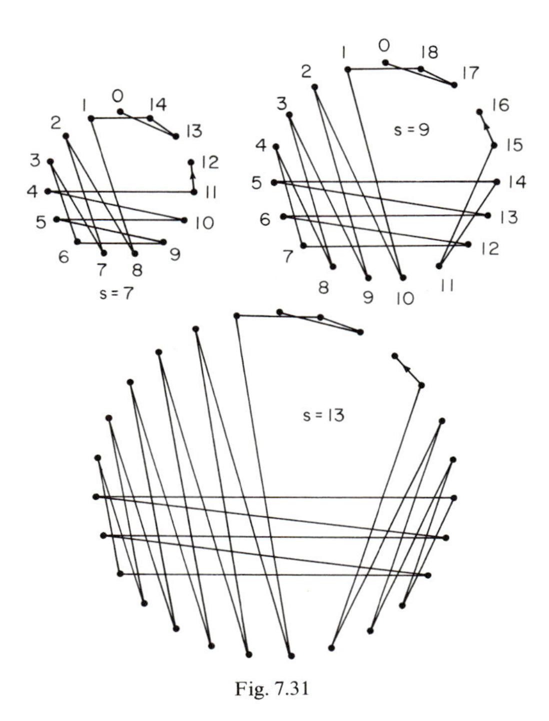

Fig. 7.31

In the general solution for $s \geqq 7$, if s is odd, the sequence of chord lengths is

$$2, 1, 2;$$
$$s, s-1, s-2, \ldots, 5, 4, 3;$$
$$s-4, s-3, s-2, s-1, s;$$
$$3, 4, 5, \ldots, s-5; 1.$$

The sequence of chords always terminates at the point $2s-2$.

Additional Adjacency Problems. Now we know how to distribute the currents in the ladder-like graph of Fig. 7.26 if $s \neq 1$ and $s \neq 3$. The current graph generates the row 0 and a scheme with rows $0, 1, 2, \ldots, 12s+5$, x, y_0 and y_1. Since the current on the arc incident with y is $6s+2$, which generates the even subgroup of $\mathbb{Z}_{12s+6}$, the letter y splits into y_0 and y_1 as in Case 2 (Section 7.4).

The graph G_{12s+9} represented by this scheme is the following: The vertices $0, 1, 2, \ldots, 12s+5$, x are all mutually adjacent, y_0 is adjacent to $0, 2, 4, \ldots$ and y_1 is adjacent to $1, 3, 5, \ldots$. There are no other arcs in the graph G_{12s+9}. The scheme represents a triangular embedding of G_{12s+9} into an orientable surface S.

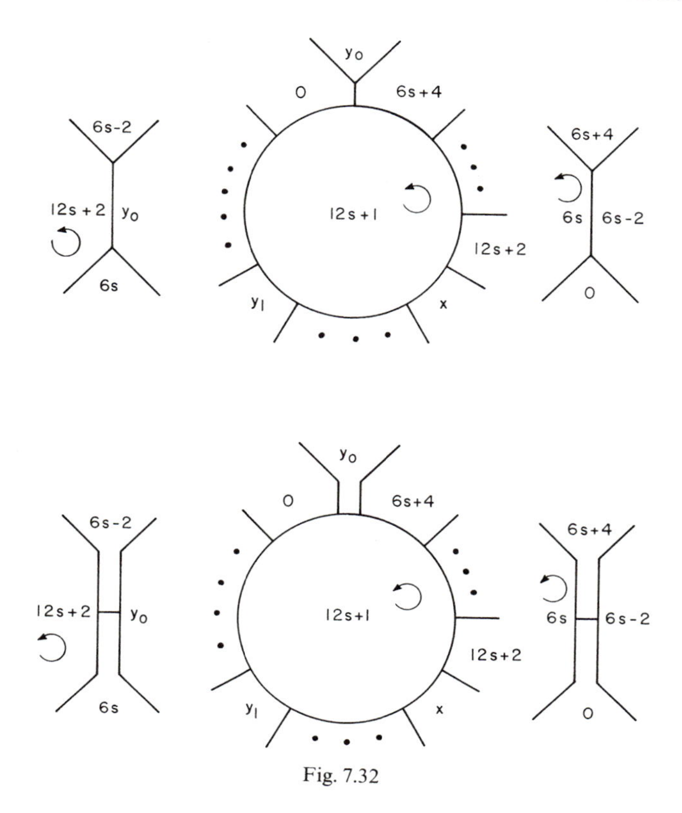

Fig. 7.32

We consider the dual map M on S. M has

$$\alpha_2 = 12s+9 \text{ countries and}$$
$$\alpha_1 = (12s+7)(6s+3)+12s+6 \text{ arcs.}$$

The genus p of S can be determined using the formula

$$\alpha_1 = 3\alpha_0 - 3E(S) = 3\alpha_0 - 6 + 6p$$

of Theorem 4.4. We obtain

$$p = 12s^2 + 9s + 1.$$

If we put $n = 12s+8$ into the right-hand side of formula (4.13), we get

$$12s^2 + 9s + 2.$$

This means we are forced to use one handle in order to manufacture the missing adjacency between x and y_0 and between y_0 and y_1. Then (4.13) will be proven for $n = 12s+8$.

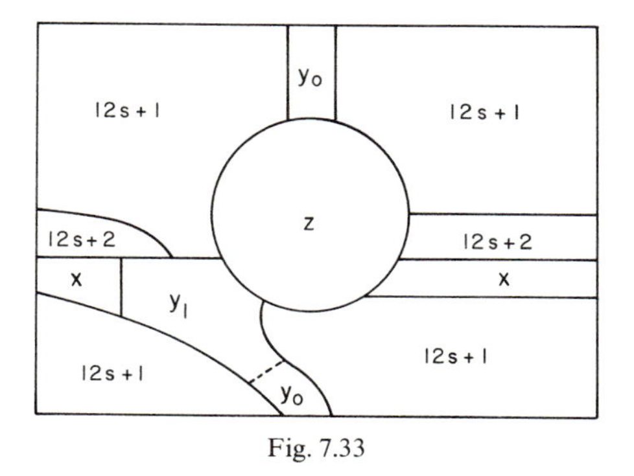

Fig. 7.33

The left side of Fig. 7.26 provides the following part of the scheme

$0.\ \ldots 6s+4,\, y_0,\, 6s+2 \ldots x, \qquad 1 \ldots 6s+9,\, 5,\, 6s+6,\, 12s+4, \qquad 4 \ldots$

$6s.\ \ldots\ .\ \ .\ \ .\ \ldots. \qquad \ldots.\ \ .\ \ . \qquad 0,\ 6s-2,\, 6s+4 \ldots$

$12s+1.\ \ldots 6s-1,\, y_1,\, 6s-3 \ldots x,\, 12s+2 \ldots 6s+4,\, 0 \qquad .\ \ .\ \ .\ \ldots$

$12s+2.\ \ldots 6s, \quad y_0,\, 6s-2 \ldots. \qquad \ldots.\ \ .\ \ .\ \ .\ \ .\ \ .\ \ldots$

Therefore there exists the partial picture of the map M shown in the top half of Fig. 7.32.

Modify the map as shown in the lower half of Fig. 7.32. We have gained adjacency $(y_0, 12s+1)$ but lost $(12s+2,\ y_0)$. The adjacencies $(0, 6s+4)$ and $(6s, 6s-2)$ are still in the map. They have merely changed their positions.

Now excise country $12s+1$ from Fig. 7.32 (lower part) and country z of the map on an extra torus given in Fig. 7.33. Identify the boundaries in the obvious way. We gain the adjacencies $(12s+2,\ y_1)$, $(x,\ y_1)$, and $(y_0,\ y_1)$.

Erase the boundary between y_0 and y_1 and call the resulting country y. Then, since we gained the adjacency $(12s+2,\ y_1)$ with the handle this accounts for the missing adjacency $(12s+2,\ y_0)$. This completes the proof for (4.13) if $n=12s+8$ and $s\neq 1$ and $s\neq 3$. A solution for $s=1$ is given in Section 5.2.

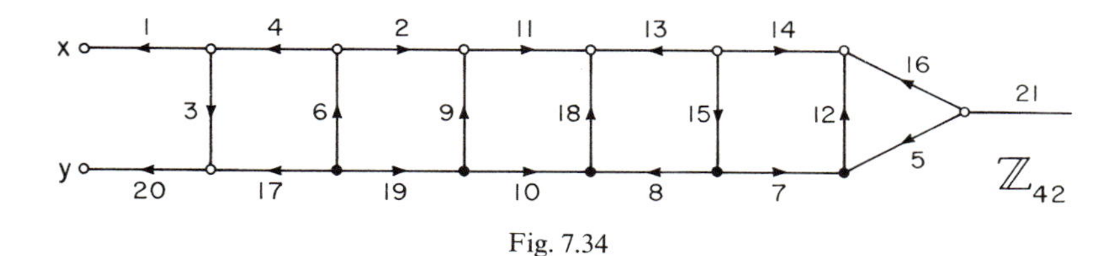

Fig. 7.34

Special Solution for $s=3$. The current graph of Fig. 7.34 has the required properties. We use the portion

0.	31	40	4	1	x	41...20	y_0	22...29	15	7...17	39...
2.						...22	y_0	24...			
3.				...	x	2...23	y_1	25...		...20	0...
13.								... 0	28	20...	
24.	13	22	28	...							

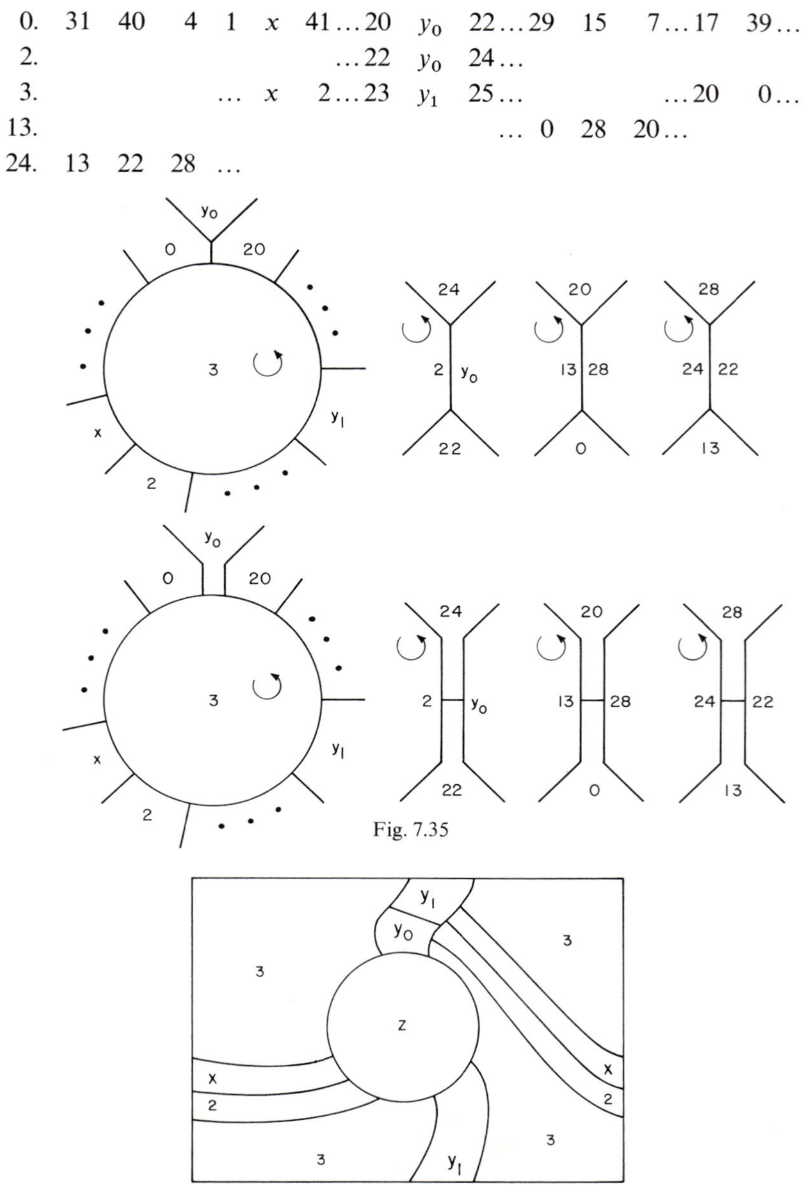

Fig. 7.35

Fig. 7.36

of the scheme to obtain the partial picture of the map as in Fig. 7.35 (upper part). Modify as shown in the lower part of Fig. 7.35. Then excise country 3 from Fig. 7.35 and country z from Fig. 7.36. Identify the boundaries to obtain a map with 44 mutually adjacent countries on the orientable surface of genus 137.

Exercises

7.6.1. Draw the ladder-like graph of Fig. 7.26 for $s=4$ and 5 and fill in all the currents according to the combs given in the text.

7.6.2. Prove that there is no solution for the aggravated chord problem (with Ch 1*) if $s=3$.

7.6.3. Find a sequence of chords with properties Ch 1), Ch 2), and Ch 3), terminating at vertex 1. Do this first for $s=2$, 3, and then for general s.

7.6.4. Find a sequence of chords with properties Ch 2) and Ch 3) where the following part somewhere in the middle is prescribed.

$$\ldots, 1, 0, s+1, s, \ldots.$$

7.6.5. Find a sequence of chords with properties Ch 2) and Ch 3) where the first and the last chords are prescribed as

$$0, s+2, \ldots, s+1, 1.$$

7.6.6. In solving the Case 11 in Section 7.2 we used associated combs. Translate these combs for $s=3, 4$, and 5 into sequences of chords. How do the properties Ch 1), Ch 2), and Ch 3) have to be modified?

7.6.7. Draw the comb associated with Fig. 7.34. Which of the properties A 1) to A 4) are satisfied?

8. Non-Orientable Cases (Index 1)

8.1. Method of Doubling

It is our intention to determine the non-orientable genus of K_n which means to prove formula (4.19). For $n \leqq 8$ this was already done in Sections 4.6 and 5.1.

First we extend the list of construction principles. The first new principle leads, together with the previous ones (Sections 2.3 and 6.1), to *orientable* triangular embeddings.

C8) *An element of order* 3 *can be the current of a dead-end-arc.* (No letter is allowed on this vertex of valence one.)

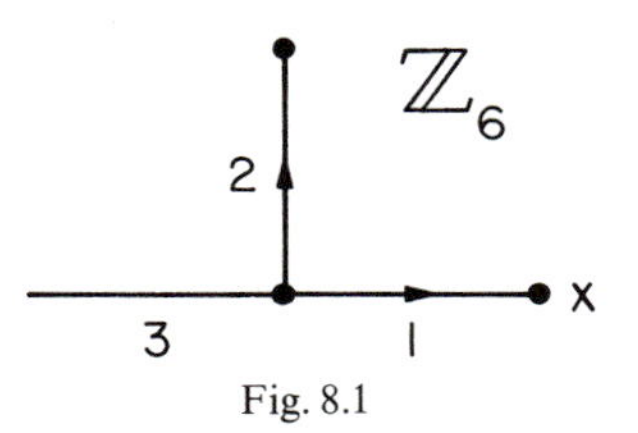

Fig. 8.1

As an example see Fig. 8.1.

Here the element 2 is of order 3. The log of this current graph reads

$$2 \quad 4 \quad 1 \quad x \quad 5 \quad 3.$$

Therefore the whole scheme is the following:

$$
\begin{array}{lcccccc}
0. & 2 & 4 & 1 & x & 5 & 3 \\
1. & 3 & 5 & 2 & x & 0 & 4 \\
2. & 4 & 0 & 3 & x & 1 & 5 \\
3. & 5 & 1 & 4 & x & 2 & 0 \\
4. & 0 & 2 & 5 & x & 3 & 1 \\
5. & 1 & 3 & 0 & x & 4 & 2 \\
x. & 0 & 1 & 2 & 3 & 4 & 5.
\end{array}
\tag{8.1}
$$

One can easily check that rule R^* is satisfied. In fact it is another representation of the well-known embedding $K_7 \lhd S_1$.

This time we have used the group $\mathbb{Z}_6$ not $\mathbb{Z}_7$. (This means that the *automorphism group* of the embedding $K_7 \lhd S_1$ has a subgroup isomorphic to $\mathbb{Z}_7$ and also a subgroup isomorphic to $\mathbb{Z}_6$.)

Observe that the current graph of Fig. 8.1 uses all the construction principles C1) to C8) except C7).

Next we prove that the new construction principle C8) together with the others always leads to a scheme satisfying Rule Δ^* (orientable!).

If d is the element of order 3 on a dead-end-arc then the row 0 reads:

$$0. \ \ldots, d, -d, \ldots.$$

Since $2d = -d$ row d reads

$$d. \ \ldots, -d, 0, \ldots.$$

This proves Rule Δ^* for row 0 as far as the dead-end-arc with current d is concerned. For the other rows it follows by the additive rule.

It is rather curious that we could not use this property C8) in any of the orientable Cases except Case 0. But soon we will need C8) for non-orientable embeddings.

In Section 2.3 we have shown that in a scheme for a graph Rule R^* and Rule Δ^* from Section 2.2 are equivalent. It is appropriate to use a similar rule which is equivalent to Rule R from Section 4.6.

By the bracket notation $i. \ldots [j\,k\,l] \ldots$ we mean that row i of the scheme considered is of the form

$$i. \ \ldots j\,k\,l \ldots \quad \text{or} \quad i. \ \ldots l\,k\,j \ldots.$$

Similarly, the notation $i. \ldots [j\,k] \ldots$ means that j and k are neighbors in row i but we are uncertain about the order. Rule R from Section 4.6 now reads as follows.

Rule *R*. *If* $i. \ldots j\,k\,l \ldots$ *then* $k. \ldots [j\,i\,l] \ldots$.

Theorem 8.1. *If a scheme (rotation) of a graph satisfies the following Rule Δ then it satisfies Rule R and vice versa.*

Rule Δ. *If* $i. \ldots j\,k \ldots$ *then* $k. \ldots [i\,j] \ldots$ *and* $j. \ldots [k\,i] \ldots$.

Proof: Assume a scheme of a graph satisfies Rule Δ. Consider a portion

$$i. \ \ldots j\,k\,l \ldots$$

of row i. Applying Rule Δ twice we get

$$k. \ \ldots [i\,j] \ldots \quad \text{and} \quad k. \ \ldots [l\,i] \ldots.$$

Since the element i appears in row k only once, it follows

$$k. \ \ldots[l\,i\,j]\ldots.$$

We leave it to the reader to prove that Rule R implies Rule Δ (Exercise 8.1.1).

Remark. If we compare Rule Δ with Rule Δ^* we could get the idea that the last part of Rule Δ might be superfluous. But it is not.

For instance the scheme

$$\begin{array}{lllll} 0. & 1 & 2 & 3 & 4 \\ 1. & 4 & 0 & 3 & 2 \\ 2. & 1 & 0 & 4 & 3 \\ 3. & 2 & 0 & 1 & 4 \\ 4. & 3 & 0 & 2 & 1 \end{array}$$

of the graph K_5 satisfies the shorter rule: *If* $i. \ldots j\,k\ldots$ *then* $k. \ldots[i\,j]\ldots$. But the scheme does not satisfy Rule R.

Now we change and extend the list of construction principles in the following way in order to get non-orientable embeddings. The property C1) we replace by

C1*) *The current graph has vertices of valence* 3, 2, *or* 1.

The following is called the *doubling principle:*

C9) *If a vertex P is of valence* 2 *then one arc incident with P must be a dead-end-arc. The current of the other arc flowing into P must be double the current of the dead-end-arc flowing out of P. (The dead-end-arc is not allowed to carry an extra letter, like x.)*

As an illustration of C9) see Fig. 8.2 and most of the following figures.

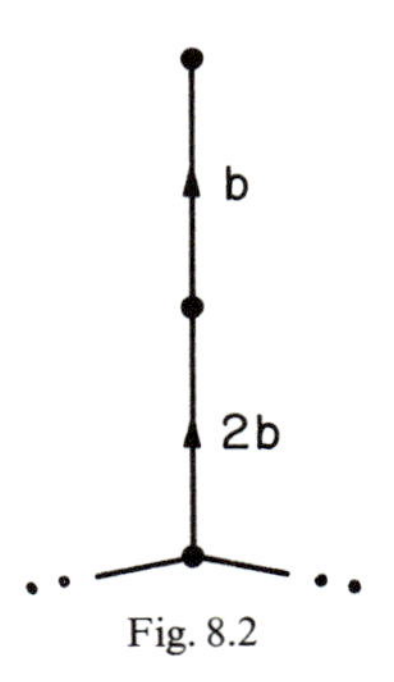

Fig. 8.2

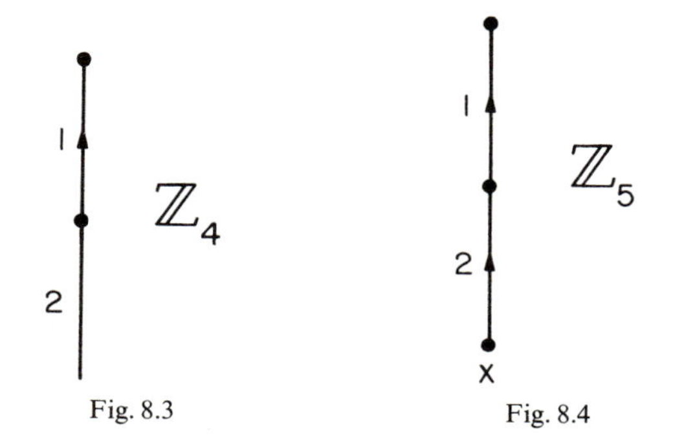

Fig. 8.3 Fig. 8.4

Let us study the row 0 and other parts of the scheme created by the partial picture in Fig. 8.2:

$$\begin{array}{rllllll}
0. & \ldots, & 2b, & b, & -b, & -2b, & \ldots \\
b. & \ldots, & ., & 2b, & 0, & -b, & \ldots \\
-b. & \ldots, & b, & 0, & -2b, & ., & \ldots \\
2b. & \ldots, & ., & ., & b, & 0, & \ldots \\
-2b. & \ldots, & 0, & -b, & ., & ., & \ldots .
\end{array}$$

We can easily check that this part of the scheme satisfies Rule Δ. Moreover, comparing rows 0 and b one sees that Rule R^* is *not* satisfied. But this does not necessarily mean that the total scheme is non-orientable.

In the example of Fig. 8.3 properties C6) from 6.1 and C9) are used. The scheme reads

$$\begin{array}{llll}
0. & 2 & 1 & 3 \\
1. & 3 & 2 & 0 \\
2. & 0 & 3 & 1 \\
3. & 1 & 0 & 2.
\end{array}$$

If we take the opposite order in rows 1 and 3, then the scheme satisfies Rule R^*. Therefore it describes an orientable triangular embedding. Of course it is $K_4 \lhd$ sphere. But this example is probably the only orientable application of property C9). The current graph of Fig. 8.4 using properties C5) and C9) generates a non-orientable embedding of K_6. In fact it is the scheme (5.3) from Section 5.1 if we replace x by 5.

Before we apply C9) let us study once again the portion of the scheme produced by Fig. 8.2:

$$\begin{array}{lrlllll}
 & 0. & \ldots, & 2b, & b, & -b, & -2b, \ldots \\
(*) & b. & \ldots, & 3b, & 2b, & 0, & -b, \ldots \\
 & 2b. & \ldots, & 4b, & 3b, & b, & 0, \ldots \\
(*) & 3b. & \ldots, & 5b, & 4b, & 2b, & b, \ldots .
\end{array}$$

Assume this scheme represents a triangular embedding of a graph into an orientable surface. Then it is possible to reverse the cyclic order of the numbers in some of the rows such that Rule R^* of Section 2.3 holds. Compare the rows 0 and b and check whether Rule R^* holds. We see that for one of them the cyclic order must be reversed. Let us reverse the order in row b (indicated by $(*)$). Then row $2b$ must stay fixed. Row $3b$ has to be reversed and so on. Therefore

$$m \cdot b \neq 0 \quad \text{for each odd } m, \tag{8.2}$$

because row 0 stays fixed and row mb must be reversed in order.

8.2. Non-Orientable Cases 0, 3, 7

Case 0. Let n be of the form $n = 12s$. We use the elements of $\mathbb{Z}_{12s-1}$ and an extra letter x to name the vertices of the complete graph K_n. Consider the current graph of Figs. 8.5, 8.6, and 8.7 (for $s=1$, $s=3$, and for general s). The ladder has $2s-2$ rungs. The vertex at the top of each rung has a clockwise rotation; the other vertex, counterclockwise. This rotation of the graph induces one single circuit (C2).

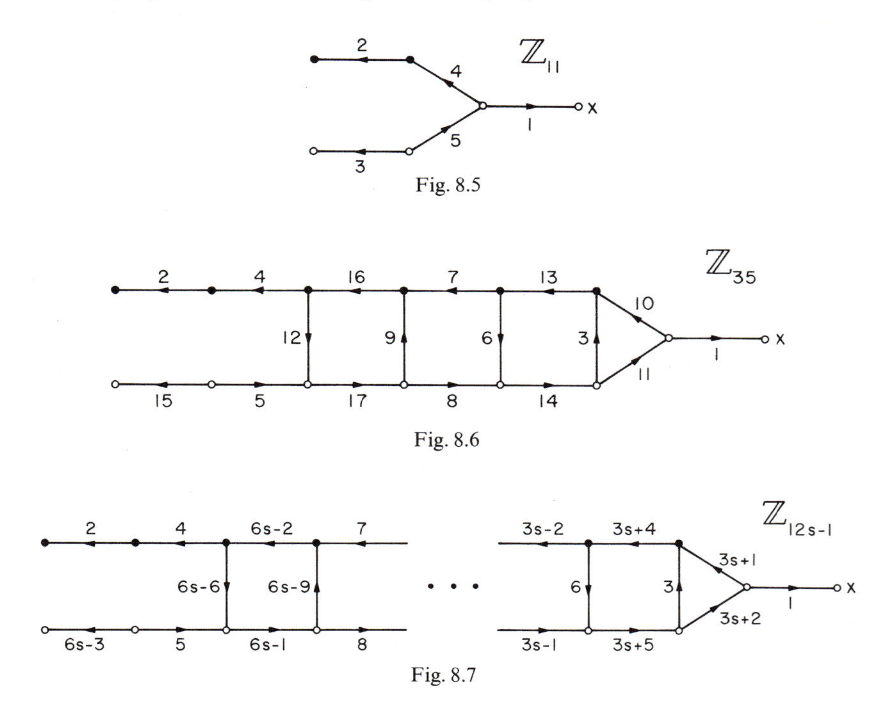

Fig. 8.7

The distribution of currents on the graph is very regular and uses each of the elements $1, 2, 3, \ldots, 6s-1$ exactly once (C3). In fact the upper (or lower) horizontals and the rungs form an arithmetic comb. At all the vertices of valence three KCL holds (C4). The current 1 on the right side of Fig. 8.7 generates the group (property C5). At the left side the property C9) is used twice; once with $b=2$ as the current on the dead-end-arc. Since

$$(12s-1)\cdot 2 \equiv 0 \pmod{12s-1}$$

is in contradiction to (8.2) we are sure that the triangular embedding of K_n given by the current graph of Fig. 8.7 is non-orientable. Therefore we have proved formula (4.19) for the non-orientable genus if $n \equiv 0 \pmod{12}$.

Case 3. Let n be of the form $n=12s+3$. Use the group $\mathbb{Z}_{12s+2}$ and an extra letter x to name the vertices of K_n. This time we use a tree as a current graph. Recall in a tree every rotation induces one single circuit. Consider the current graph shown in Fig. 8.10. Figs. 8.8 and 8.9 illustrate examples for $s=1$ and 2. Again the distribution of the elements $1, 2, \ldots, 6s+1$ is very regular. The lower verticals (except the first one) and the horizontals (except the first one) form a very easy arithmetic comb. Notice that the element $6s+1$ is of order 2 and is the current of a dead-end-arc (property C6). At all the vertices of valence two the doubling property C9) holds.

It is easy to see that the embedding produced by the current graph of Fig. 8.10 is non-orientable: take the current $b=4s+2$ on the right-

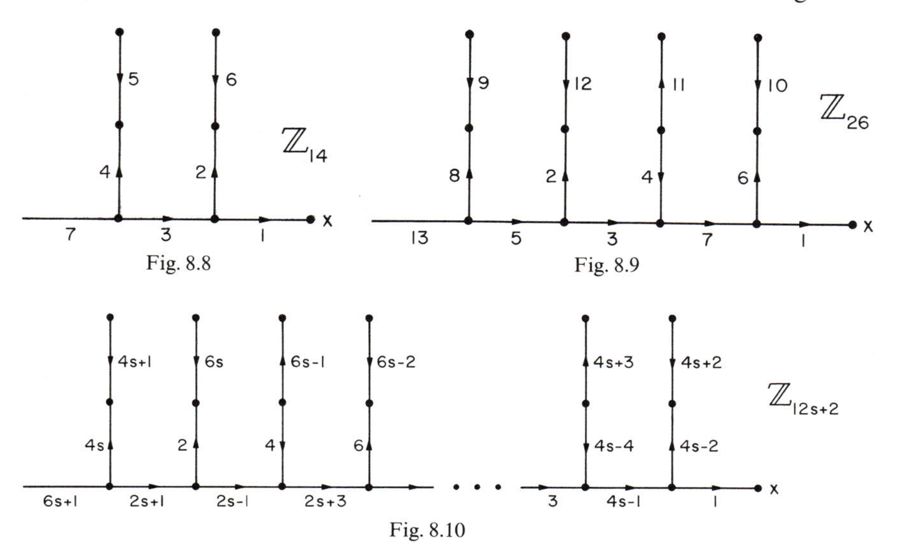

Fig. 8.8

Fig. 8.9

Fig. 8.10

most vertical dead-end-arc. Since

$$(6s+1)\,b \equiv (6s+1)(4s+2) \equiv 0 \pmod{12s+2}$$

contradicts formula (8.2), the embedding cannot be orientable. Thus formula (4.19) is proved for $n \equiv 3 \pmod{12}$.

Case 7. Let n be of the form $n = 12s+7$ and $s \geqq 1$. We use the elements of the group $\mathbb{Z}_{12s+6}$ and an extra letter x to identify the vertices of K_n. Consider the current graph of Fig. 8.12 (Fig. 8.11 in the special case $s=1$).

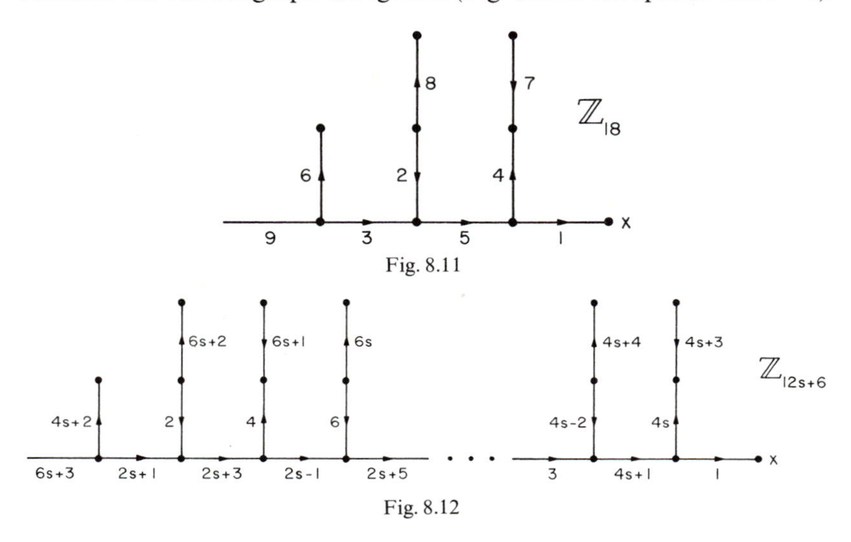

Fig. 8.11

Fig. 8.12

It is again a very easy arithmetic comb. Notice that at the left side of the current graph the elements $4s+2$ and $6s+3$ are of order 3 and 2 respectively. So we use the construction principles C8) and C6).

Of course each of the elements $1, 2, \ldots, 6s+3$ is used once and KCL holds at every vertex of valence three. Moreover the doubling property C9) holds at every vertex of valence 2.

At the right side of Fig. 8.12 we see the current $4s+4$. Compare this with Fig. 8.2. The doubling property is used with $b = 4s+4$. Since

$$(6s+3)(4s+4) \equiv 0 \pmod{12s+6}$$

does not agree with inequality (8.2) the scheme produced by the graph of Fig. 8.12 is non-orientable.

Exercises

8.2.1. Prove the second part of Theorem 8.1: If in a scheme for a graph Rule R is satisfied then Rule Δ holds too.

8.2.2. Construct a triangular embedding of K_{13} into a non-orientable surface using the group $\mathbb{Z}_{13}$ and the doubling property C9) three times.

8.2.3. Construct a non-orientable triangular embedding of K_{10}. Use the group $\mathbb{Z}_9$, doubling, and property C8).

8.2.4. Find a non-orientable triangular embedding of K_{16}. Use $\mathbb{Z}_{15}$, property C8), and the doubling property C9) twice.

8.3. Cascades

We would like to introduce another very helpful construction principle which can be used for non-orientable embeddings.

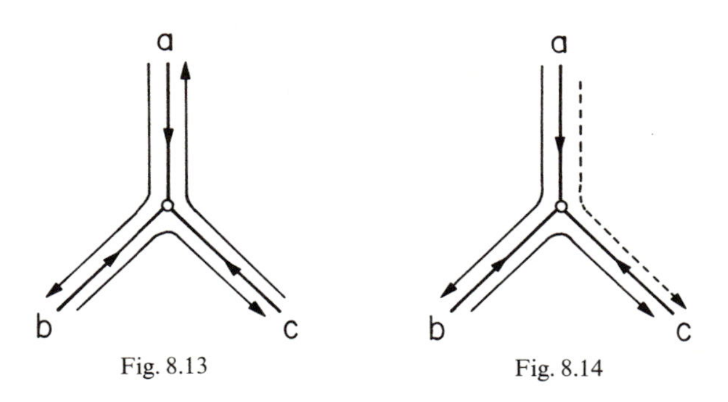

Fig. 8.13 Fig. 8.14

Consider a vertex of valence 3 in a current graph (Fig. 8.13) with the inward flowing currents a, b, c satisfying KCL, namely, $a+b+c=0$. The log of the circuit produces row 0 which reads

$$0. \quad \ldots, a, \ -b, \ldots, b, \ -c, \ldots, c, \ -a, \ldots$$

As we know this agrees exactly with Rule Δ^*. But this time we only need the weaker Rule Δ. For instance, let us reverse the order of the last pair:

$$0. \quad \ldots, a, \ -b, \ldots, b, \ -c, \ldots, -a, \ c, \ldots$$

This agrees with Rule Δ as far as the three elements a, b, c are concerned. In Fig. 8.14 we try to visualize this possibility. *Along the dashed line* indicating part of the circuit one has to do everything in *reverse:* If the current is supposed to be a, record $-a$ instead of a. If the rotation is clockwise, consider it as counterclockwise and vice versa.

Now we need a place where we can connect the dashed line with the normal line indicating the journey (circuit). This will be at the middle of an arc as illustrated in Fig. 8.15. But this requires a change in the orientation of "part of the arc". All these ideas finally lead to the following definition.

A *cascade* is a current graph with the following additional modifications:

1) Some of the arcs are called *broken arcs*. A broken arc is divided by its *midpoint* into two *half arcs*. (In order to avoid confusion with property C8) and others we do not consider the midpoints as vertices.)

2) The broken arcs carry currents as the other arcs do.

3) The half arcs of a broken arc are both oriented toward the midpoint or both away from the midpoint.

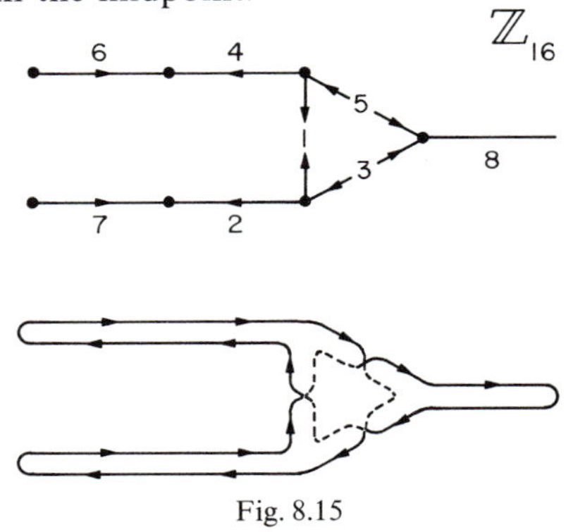

Fig. 8.15

For instance the cascade of Fig. 8.15 has three broken arcs.

In a current graph we considered the circuit (or circuits) which is induced by the rotation of the current graph. (We have interpreted the circuit as a trip of a traveller.) For cascades we have to modify such a circuit (or trip) as follows:

Begin the trip at an unbroken arc and act as in a current graph. But at the moment you pass the midpoint of a broken arc "multiply your behavior by -1". By this we mean two things:

1) If you approach a counterclockwise vertex act in a clockwise manner and vice versa.

2) If you are on an arc with current a and you use the arc in the same direction as the arc is oriented then record in the log of the trip $-a$ instead of a, and if the direction is opposite record $+a$.

Of course when you pass a midpoint of a broken arc the second time then you have to act normally again.

The parts of the trip where you have to do everything wrong (multiply by -1) are indicated by a dashed line in the lower half of Fig. 8.15.

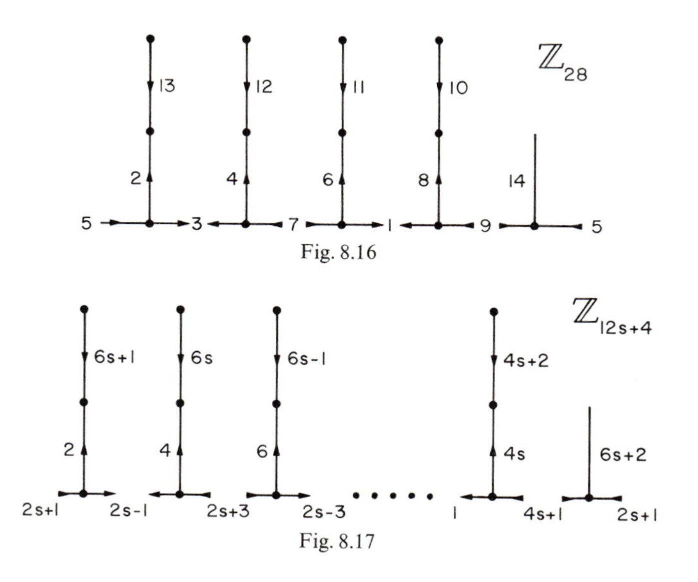

Fig. 8.16

Fig. 8.17

The log of the trip in this example determines row 0 as follows.

0. 4 10 6 12 11 3 2 9 7 14 1 5 8 13 15.

Using the additive rule one obtains a scheme for K_{16} satisfying Rule R.

Case 4. It is very easy to generalize this example for $n=12s+4$ if $s \geqq 1$. Use the elements of $\mathbb{Z}_{12s+4}$ to name the vertices of K_{12s+4}. Consider the cascade of Fig. 8.17. Fig. 8.16 shows the example $s=2$. The half arc at the right carrying current $2s+1$ and the half arc at the left with the same current have to be considered as *one* broken arc. It is easy to check that each of the elements $1, 2, \ldots, 6s+2$ is used exactly once, the last being on a dead-end-arc. KCL holds at each vertex of valence 3. At all the vertices of valence 2 property C9) holds.

The log of the trip induced by the cascade of Fig. 8.17 contains each element of $\mathbb{Z}_{12s+4}$ which is $\neq 0$. Of course this has to be checked. Since the local picture around each vertex of valence 3 is of the form of Fig. 8.14 and since all the other requirements are satisfied, the scheme we obtain satisfies Rule Δ. The scheme defines a triangular embedding of K_{12s+4} into a closed surface. We can prove that the surface is non-orientable in the following way.

Consider the following part of the scheme.

$$\begin{array}{rllllll}
0. & \ldots 4s, & 8s+2, 4s+2, 8s+4 \ldots 2, & 6s+3, & 6s+1, & 12s+2 \\
2. & \ldots 4s+2, 8s+4, & . \quad . \quad \ldots . \quad . & 6s+3, & & 0 \\
4s+2. & \ldots \quad . & 0, 8s+4, \quad 2 \ldots . & . & . & .
\end{array}$$

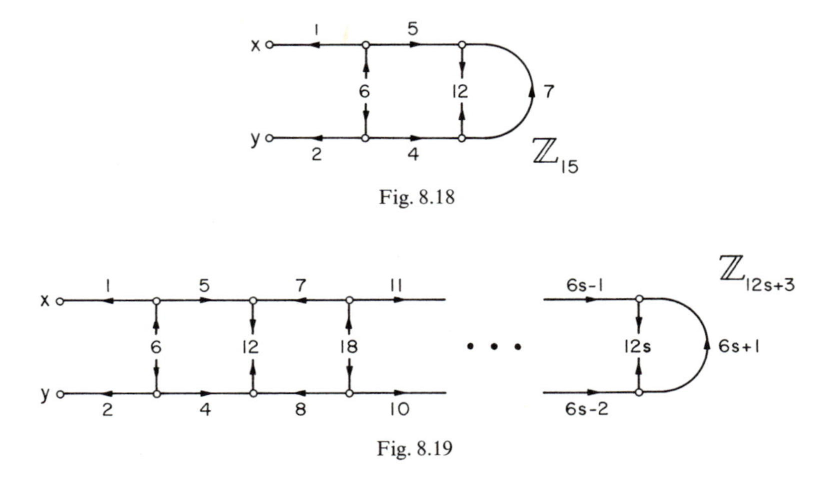

Fig. 8.18

Fig. 8.19

No matter which of these three rows are reversed, rule R^*, or the equivalent rule Δ^*, cannot hold for *all* three. Hence the surface is non-orientable.

Case 5. Let n be of the form $n=12s+5$. Use the group $\mathbb{Z}_{12s+3}$ and the cascade of Fig. 8.19. Fig. 8.18 shows an example for $s=1$. Notice that the numbers

(8.3) $$1, 4, 7, \ldots, 6s-2$$

appear alternating on the upper and lower horizontals. The numbers

(8.4) $$2, 5, 8, \ldots, 6s-1$$

are also alternating on the lower and upper horizontals. The broken rungs carry the currents

(8.5) $$6, 12, \ldots, 12s$$

in this order. The current $6s+1$ from the right-most arc, the elements of the collections (8.3), (8.4), and (8.5), and all their negatives exhaust the non-zero elements of $\mathbb{Z}_{12s+3}$.

At every vertex of valence 3 KCL is true. The currents of the two dead-end-arcs are 1 and 2. Both are generators of the group $\mathbb{Z}_{12s+3}$. It is easy to check that the cascade induces one single trip whose log contains each element of $\mathbb{Z}_{12s+3}$ and x and y. Therefore we have found a triangular embedding of $K_{12s+5}-K_2$ into a closed surface. The missing adjacency between x and y can be manufactured by adding one cross cap as shown at the end of Section 5.3. The resulting surface is non-orientable.

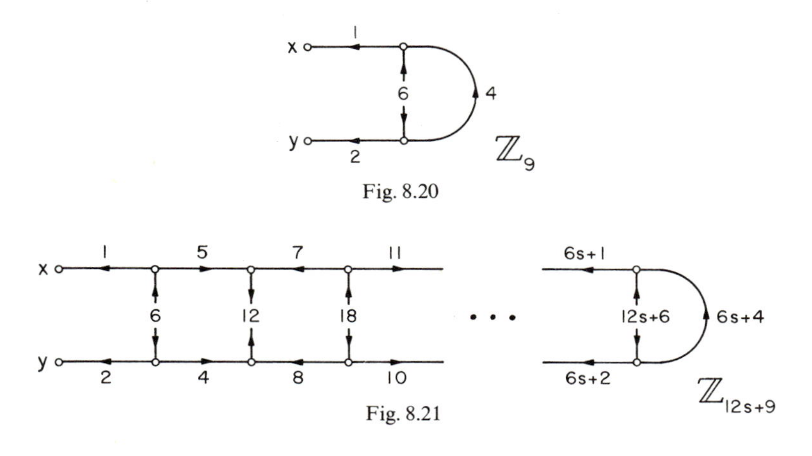

Fig. 8.20

Fig. 8.21

Case 11. If $n=12s+11$ we use the cascade of Fig. 8.21 where the currents are from $\mathbb{Z}_{12s+9}$. Fig. 8.20 illustrates the example $s=0$. Notice that we have used the cascade of Fig. 8.20 to obtain the scheme (5.8) in Section 5.3. The right part of Fig. 8.21 is practically the same as in Case 5. The cascade shows that there exists a triangular embedding of $K_{12s+11}-K_2$ into a closed surface for $s=0, 1, \ldots$.

Remarks About Case 1. Let us consider some examples where $n\equiv 1$ (mod 12). The current graph of Fig. 8.22 provides a triangular embedding of K_{13} into a non-orientable surface. For the next case $n=25$, Figs. 8.23,

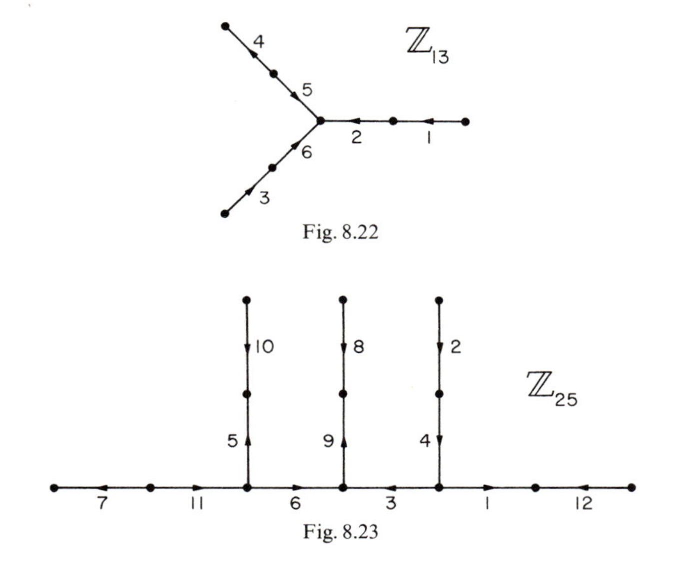

Fig. 8.22

Fig. 8.23

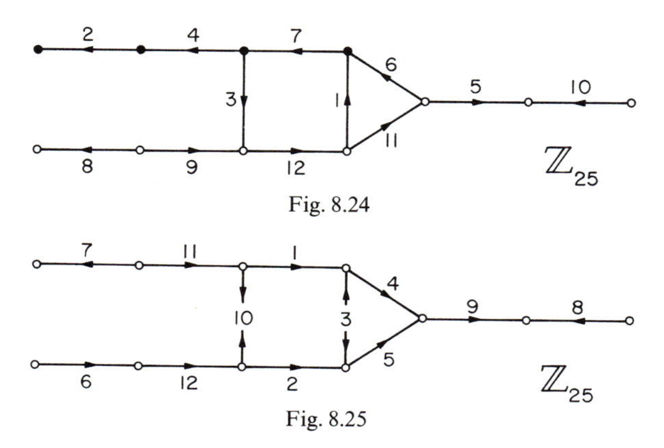

Fig. 8.24

Fig. 8.25

8.24 and 8.25 show three different solutions for an embedding $K_{25} \lhd S$. All three can be considered as generalizations of the example of Fig. 8.22.

Landesman and Youngs constructed in the paper [49] a solution for all $n \equiv 1 \pmod{12}$ generalizing Fig. 8.24 and a solution generalizing Fig. 8.25. In spite of the title of this paper both solutions are very difficult.

In this book we will give an easier solution which uses other methods. (See Section 10.3.) Landesman has also tried to find a solution using a simple comb as in Fig. 8.23. He found comb solutions for $n = 25$, 37, 49, 61, 73, and 85. However he had no success in finding a general pattern.

Case 8. Let n be of the form $n = 12s + 8$. As in the orientable Case 8 we first construct a certain map M with countries $1, 2, \ldots, 12s+6, x, y_0, y_1$ on a certain orientable surface S. It will turn out that the additional adjacency problem is much easier than in the orientable Case 8. Therefore

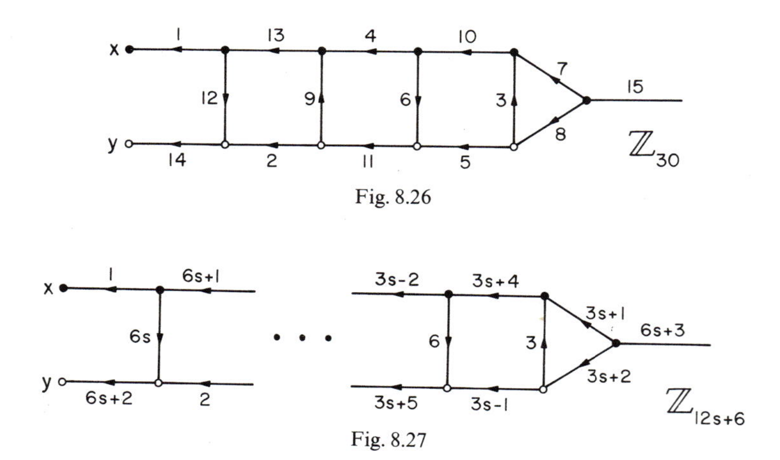

Fig. 8.26

Fig. 8.27

we use a very simple current graph, shown by Fig. 8.27. For $s=2$ see Fig. 8.26. The method works also for $s=0$, where Fig. 8.27 is reduced to only three arcs. The current graph of Fig. 8.27 defines a map M on the surface S. As we have already calculated in Section 7.6 the genus of S equals $12s^2+9s+1$. Therefore

$$E(S) = -24s^2 - 18s. \tag{8.6}$$

In the map M the country x must be adjacent to an even numbered country α and an odd numbered country β with α and β adjacent to each other, as shown in Fig. 8.28. Moreover, the country y_0 must be

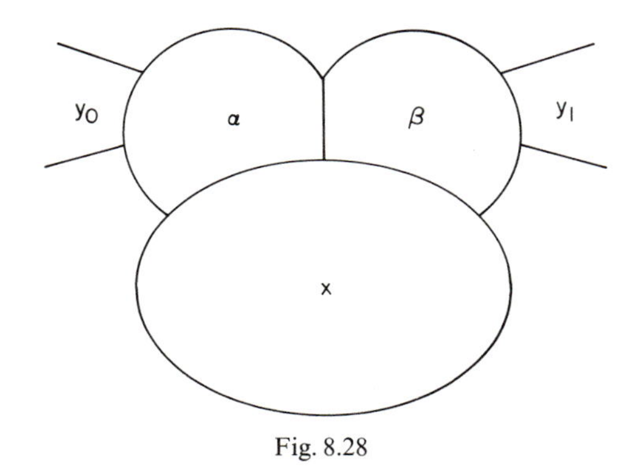

Fig. 8.28

adjacent to α and y_1 to β. Excise the interior of a closed 2-cell inside country α and identify opposite points on the boundary. Do the same for β. This construction adds two cross caps to S giving us the new surface S'. Therefore

$$E(S') = E(S) - 2. \tag{8.7}$$

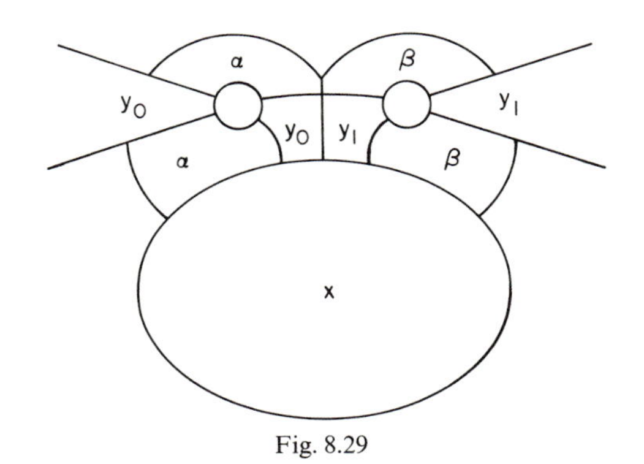

Fig. 8.29

Now enlarge countries y_0 and y_1 as shown in Fig. 8.29. (Use compositions and subdivisions of dimension 1 or 2.) Erase the boundary between y_0 and y_1 and call this newly created country y. The map on S' has n countries each adjacent to all the others. S' is non-orientable. The genus q of S' can be computed by (8.6), (8.7) and $E(S')=2-q$. We obtain

$$q=24s^2+18s+4.$$

This completes the proof of formula (4.19) for $n=12s+8$.

8.4. Orientable Application

It may come as a surprise that cascades can even be used for constructing orientable triangular embeddings. M. Jungerman [41] found the following solution for half of Case 9.

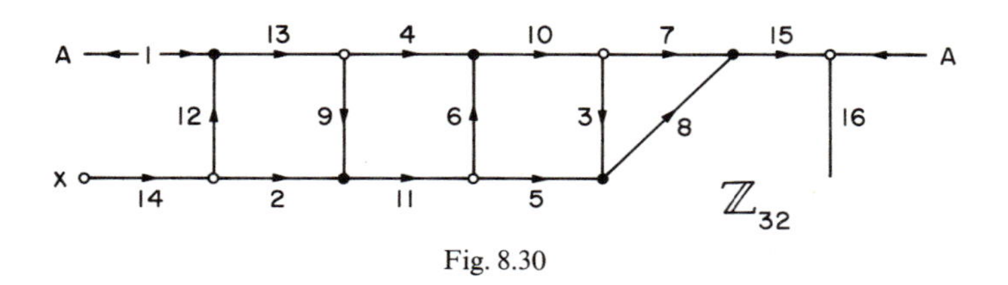

Fig. 8.30

The Figs. 8.30 and 8.31 are cascades with only one broken arc. (The two points A are to be identified.) One can check that the given rotation induces one single circuit and its log contains each group

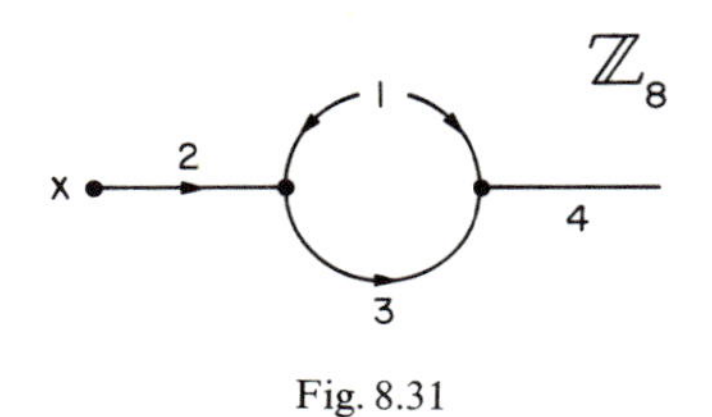

Fig. 8.31

element $\neq 0$ exactly once. KCL holds at each vertex of valence 3. The element of order 2 occurs on a dead-end-arc. The other dead-end-arc carries the current 14 or 2 respectively, which generates the even subgroup. This means in the scheme produced the row x has two cycles.

The scheme produced by the cascade of Fig. 8.31 reads as follows:

$$\begin{array}{lllllllll}
0. & 4 & 5 & 6 & x & 2 & 7 & 3 & 1 \\
1. & 5 & 6 & 7 & x & 3 & 0 & 4 & 2 \\
2. & 6 & 7 & 0 & x & 4 & 1 & 5 & 3 \\
3. & 7 & 0 & 1 & x & 5 & 2 & 6 & 4 \\
4. & 0 & 1 & 2 & x & 6 & 3 & 7 & 5 \\
5. & 1 & 2 & 3 & x & 7 & 4 & 0 & 6 \\
6. & 2 & 3 & 4 & x & 0 & 5 & 1 & 7 \\
7. & 3 & 4 & 5 & x & 1 & 6 & 2 & 0 \\
x. & (6 & 4 & 2 & 0) & (7 & 5 & 3 & 1).
\end{array}$$

From the construction principles used we know that this scheme satisfies Rule R. But if we reverse the order in all the odd numbered rows, it turns out that Rule R^* holds everywhere. In fact we then get the scheme ($n=9$) already displayed in Section 5.2 where we interpreted the row x as a cylinder.

The generalization for $n=12s+9$ where s is even using the group $\mathbb{Z}_{12s+8}$ is very easy. Fig. 8.30 shows the next case $s=2$. In general there is an arithmetic comb where the verticals carry the currents

$$6s, \ldots, 9, 6, 3.$$

The orientations alternate. Then the upper horizontal at the left is the only broken arc in the cascade and carries the current 1. Everything else is uniquely determined by KCL. The rotations at the vertices of the upper (lower) horizontals alternate clockwise and counterclockwise (counterclockwise and clockwise).

The cascade has the following two additional properties:

F 1) *The group is of even order.*

F 2) *If the current of an arc (broken or not) is odd then the arc appears in the circuit twice in the same direction. If the current is even then the circuit uses the arc in both directions.*

Of course if an arc is used twice in the same direction the circuit must have the "wrong behavior" one of those times.

It is very easy to show that the two properties F 1) and F 2) guarantee that the produced scheme is orientable. Simply reverse the order of the odd-numbered rows in the scheme.

We omit the details here because in Section 6.5 we already presented a solution for Case 9 for all s.

9. Solutions of Index 2 and 3

9.1. Examples and Method

Each of the current graphs we used in the previous chapters has the property that the given rotation induces *one* single circuit. The log of the circuit provides row 0 and all the other rows are determined by the additive rule (or for the extra rows (as row x, etc.) by Rule R^*). We say row 0 *generates* the whole scheme and the solution is of *index* 1.

But we can also use more than one generating row. The scheme for $n=14$ in Section 5.2 has the property that the even (odd) numbered rows are obtained from row 0 (1) by the additive rule. Here we have two generating rows: row 0 and row 1. We say the scheme is of *index* 2.

The scheme for $n=12$ in Section 5.2 is of index 4, because there are four generating rows. For the three orientable Cases $n \equiv 3$, 5, and 6 (mod 12) we will give an index 3 solution soon.

A scheme or triangular embedding of higher index can also be obtained from a current graph. We will consider only solutions of index 2 and 3. As examples for index 2 consider the current graphs of Figs. 9.1, 9.2 and 9.3. They all have the following pertinent properties:

D1) *Each vertex has valence* 3, 2 *or* 1.

D2) *The given rotation induces two circuits.*

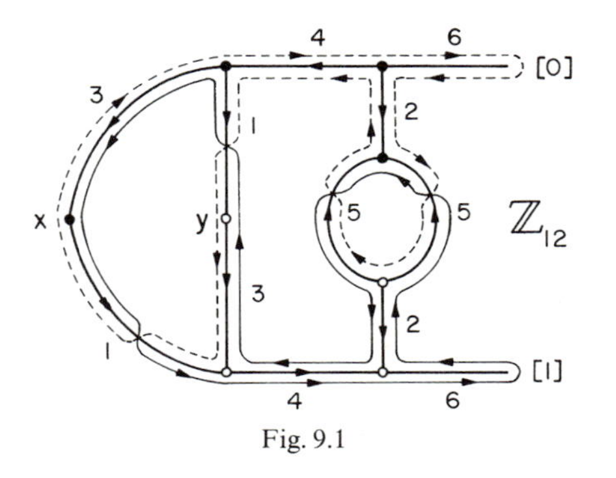

Fig. 9.1

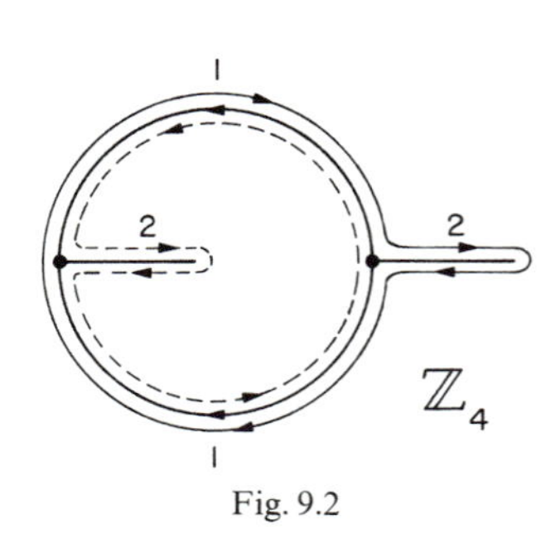

Fig. 9.2

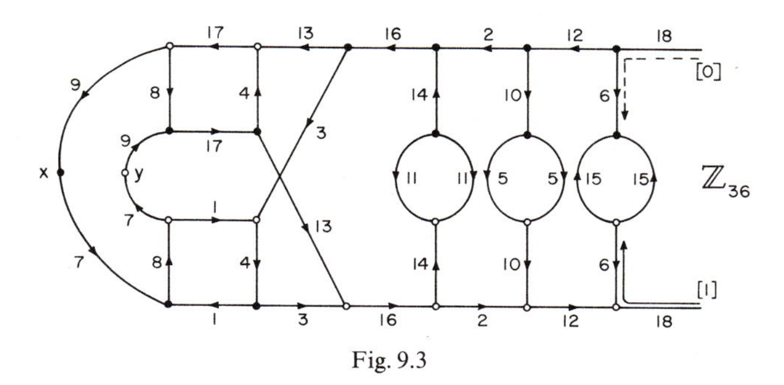

Fig. 9.3

D3) *Each element of the group, except* 0, *appears exactly once in the log of each circuit.*

D4) KCL *is true at each vertex of valence* 3.

D5) *Each vertex of valence* 2 *is identified by a symbol like* $x, y, \ldots$. *The currents of the two arcs incident with this vertex are odd numbers. The sum of both currents (counted as inward flowing) generates the subgroup of all even elements.*

D6) *An element of order* 2 *may be a current of a dead-end-arc.*

D7) *If an arc is used in both circuits then its current is an odd number. If an arc is used twice in the same circuit then its current is an even number.*

The first properties D1) to D6) are more-or-less obvious modifications of the construction principles C1) to C6) for index 1 solutions. Notice that D6) is weaker than C6). In Fig. 10.7 we will see an example of a current graph in which the element of order 2 is not on a dead-end-arc. Property D7) is an additional one which is very important. Now we explain how a current graph with properties D1) to D7) creates an orientable triangular embedding of a graph.

We write down the log of both circuits, one of them denoted by [0], the other by [1]. In the example of Fig. 9.1 we obtain:

$$\begin{array}{lccccccccccccc} [0]. & 6 & 2 & 7 & 5 & 10 & 4 & 1 & y & 3 & 11 & x & 9 & 8 \\ [1]. & 6 & 10 & 5 & 7 & 2 & 8 & 9 & y & 11 & 3 & x & 1 & 4. \end{array}$$

The log [0] will give immediately the row 0:

$$0.\quad 6\quad 2\quad 7\quad 5\quad 10\quad 4\quad 1\quad y\quad 3\quad 11\quad x\quad 9\quad 8.$$

The row 1 we obtain from [1] by adding +1 to each element without changing the order:

$$1.\quad 7\quad 11\quad 6\quad 8\quad 3\quad 9\quad 10\quad y\quad 0\quad 4\quad x\quad 2\quad 5.$$

Then each even (odd) numbered row is obtained by the additive rule from row 0 (from row 1). The rows x and y are determined by Rule R^*. The total scheme of this example is given in Section 5.2, and gives an embedding $K_{14}-K_2 \lhd S_9$. In the same way one obtains from Fig. 9.3 an embedding $K_{38}-K_2 \lhd S_{99}$. Ringel and Youngs [77] gave a general proof for the existence of a triangular embedding of $K_{12s+2}-K_2$ into an orientable surface, but only for odd numbers s.

Now we will prove that the properties D 1) to D 7) always guarantee that the scheme obtained satisfies Rule Δ^*.

Assume that part of row i is

$$i. \quad \ldots j, k \ldots .$$

Therefore the log of one of the two circuits reads

$$[a]. \quad \ldots j-i, k-i \ldots$$

where $a=0$ or 1 and by the kind of construction

$$a \equiv i \pmod 2 \tag{9.1}$$

holds. Consequently somewhere in the current graph the partial picture of Fig. 9.4 appears and we see that the log of circuit $[b]$ has the form

$$[b]. \quad \ldots i-k, j-k \ldots$$

because $h=j-k$ by KCL. Property D 7) says that $k-i \equiv b-a \pmod 2$. From (9.1) it follows

$$k \equiv b \pmod 2.$$

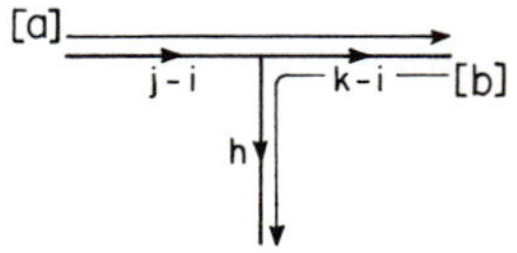

Fig. 9.4

Therefore row k has the form

$$k. \quad \ldots i, j. \quad \ldots .$$

This shows that Rule Δ^* holds. This proof includes the possibility that $k-i$ may be of order 2.

It is also easy to prove that D 5) guarantees that the row x, manufactured by Rule R^*, will be a full permutation of all elements of the group. We leave this proof to the reader.

The example of Fig. 9.2 describes an embedding of K_4 into the sphere. A generalization of Fig. 9.2 is not known.

Example of an index 3 *solution.* To introduce the ideas of index 3 solutions consider the current graph of Fig. 9.5 where the two ends A (and also the two ends B) are identified. Identification of the ends of a ladder-like graph with such a twist yields a *Möbius-type* ladder-like graph. Without such a twist we obtain a *cylindrical-type* ladder.

The following properties E 1) to E 6) are the construction principles for solutions of index 3. They are valid in the example considered in Fig. 9.5. Property E 5) does not apply in this example.

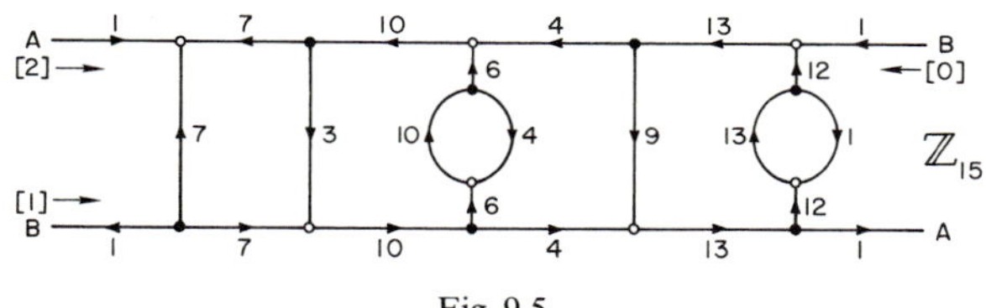

Fig. 9.5

E 1) *Each vertex is of valence* 3.

E 2) *The given rotation induces three circuits, labeled* [0], [1], [2].

E 3) *Each element of the group, except* 0, *appears exactly once in the log of each circuit.*

E 4) *With the exception of the vertices identified by letters, KCL holds at each vertex of valence three.*

E 5) *Each vertex identified by a letter has the following two properties:*

(1) *Each of the three circuits passes through the vertex.*

(2) *The sum of the three currents (counted as inward flowing) at the vertex generates the subgroup of all multiples of* 3.

E 6) *If an oriented arc with current c is in the circuit* $[a]$, *and the same arc but with opposite orientation and current* $-c$ *belongs to the circuit* $[b]$, *then* $c \equiv b - a$ (mod 3). (The situation is shown in Fig. 9.7.)

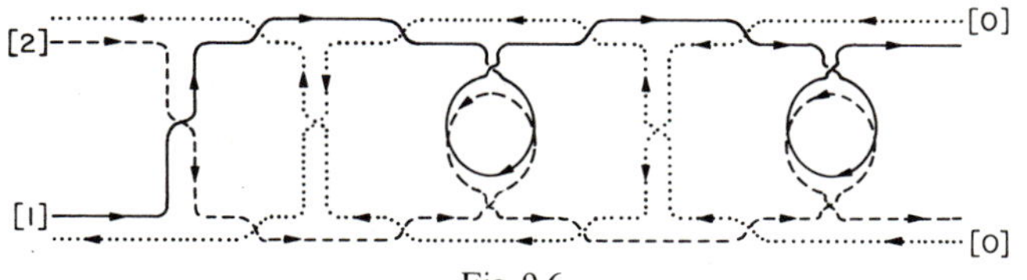

Fig. 9.6

Fig. 9.7

The three circuits are displayed in Fig. 9.6. The reader is invited to redraw the graph and trace out the circuits color coding them by using three different colors.

If the local picture of Fig. 9.7 is true, we say *that circuit* [a] *meets circuit* [b] *along the arc with current* c.

The property E 6) is a very strong requirement and makes it hard to discover an index 3 solution. It means for instance that all the arcs along which circuit [a] meets itself have currents $\equiv 0 \pmod 3$.

For each of the three circuits given by Figs. 9.5 and 9.6 we write down the logs:

[0].	1	13	9	11	5	12	7	14	2	6	4	10	3	8
[1].	14	7	8	5	9	4	10	6	11	2	3	1	13	12
[2].	1	8	7	10	6	11	5	9	4	13	12	14	2	3.

For an index 3 solution the additive rule has to be modified in the following way.

For each element $i \equiv a \pmod 3$ *the row* i *is obtained from the log of circuit* [a] *by adding* $+i$ *to the elements of the row and making no change in the order* ($a = 0$, 1 or 2).

We thus obtain the scheme

0.	1	13	9	11	5	12	7	14	2	6	4	10	3	8
3.	4	1	12	14	8	0	10	2	5	9	7	13	6	11
6.	7	4	0	2	11	3	13	5	8	12	10	1	9	14
9.	10	7	3	5	14	6	1	8	11	0	13	4	12	2
12.	13	10	6	8	2	9	4	11	14	3	1	7	0	5
1.	0	8	9	6	10	5	11	7	12	3	4	2	14	13
4.	3	11	12	9	13	8	14	10	0	6	7	5	2	1
7.	6	14	0	12	1	11	2	13	3	9	10	8	5	4
10.	9	2	3	0	4	14	5	1	6	12	13	11	8	7
13.	12	5	6	3	7	2	8	4	9	0	1	14	11	10
2.	3	10	9	12	8	13	7	11	6	0	14	1	4	5
5.	6	13	12	0	11	1	10	14	9	3	2	4	7	8
8.	9	1	0	3	14	4	13	2	12	6	5	7	10	11
11.	12	4	3	6	2	7	1	5	0	9	8	10	13	14
14.	0	7	6	9	5	10	4	8	3	12	11	13	1	2.

We could easily check that in this scheme for K_{15} rule Δ^* holds. But it is better to prove in general that property E 5) together with the modified additive rule guarantees that rule Δ^* holds: Assume the portion

$$i. \ \dots \ j, k \ \dots$$

appears in the scheme. It follows that in one of the three circuits, say $[a]$, we have

$$[a]. \ \dots j-i, k-i \dots \quad \text{and} \quad a \equiv i \pmod 3. \tag{9.2}$$

Along the arc with current $k-i$ the circuit $[a]$ meets another circuit, say $[b]$. The possibility $[a]=[b]$ is not exluded.

From Fig. 9.4 we see that the log of circuit $[b]$ has the form

$$[b]. \ \dots i-k, h \dots$$

where $h=j-k$. Property E 6) says that $k-i \equiv b-a \pmod 3$. From (9.2) it follows that

$$k \equiv b \pmod 3.$$

By the modified additive rule we get

$$k. \ \dots i, j \dots$$

establishing rule Δ^*.

9.2. Orientable Cases 3 and 5

Case 3. If n is of the form $n=12s+3$ we will be able to show that there exists a triangular embedding of K_n into an orientable surface. As we have seen in the previous section the current graph of Fig. 9.5 provides a triangular embedding of K_{15}. The generalization is easy. First see the next case $n=27$ as illustrated in Fig. 9.8. In both examples the index 3 construction principles E 1) to E 4) and E 6) are satisfied.

Moreover observe that in the ladder-like graph of both examples the 3rd, 5th, ... rung is subdivided by two vertices and a lune which is called a *globe*. The two curved arcs forming the globe are called *globular arcs*.

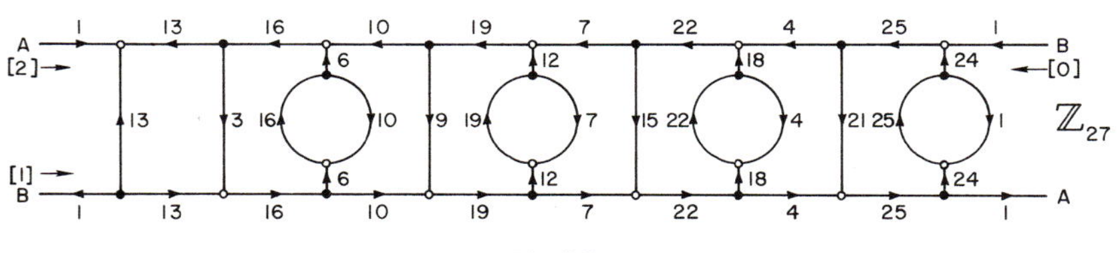

Fig. 9.8

Notice that any two corresponding horizontals and the globular arc between them always carry the same current. With this in mind it is easy to complete the unfinished current graph of Fig. 9.9 in the general Case $n=12s+3$. The crucial part of Fig. 9.9 is just a simple arithmetic comb. Of course one has to check that the properties E 1) to E 4) and E 6) are satisfied.

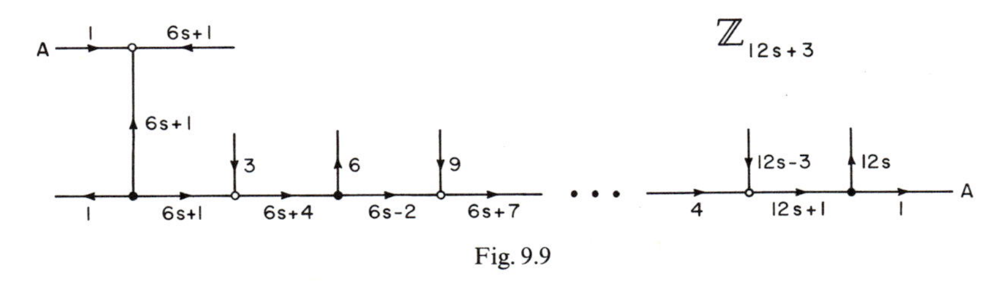

Fig. 9.9

Remark. These kinds of solutions are very touchy. For instance if you interchange the two currents 10 and 4 (including the direction) in the first globe in Fig. 9.5, property E 3) would be violated.

Case 5. The technique developed in solving Case 3 makes the solution to Case 5 quite simple. We will construct a triangular embedding of $K_n - K_2$ into an orientable surface for all $n \equiv 5 \pmod{12}$. Let n be of the form $n=12s+5$. We use the symbols x and y to designate the two exceptional vertices and the elements of the group $\mathbb{Z}_{12s+3}$ for the other vertices.

The example $s=1$ is illustrated in Fig. 9.10. There are two vertices x and y. Observe that all the properties E 1) to E 6) are satisfied.

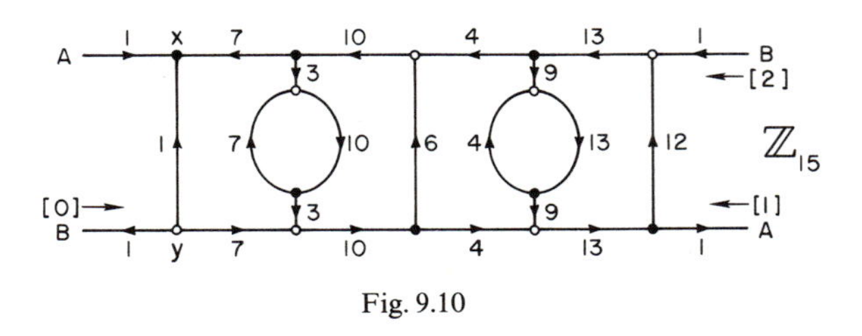

Fig. 9.10

The logs of the three circuits read

[0]. 1 x 8 5 9 4 13 12 14 y 7 10 6 11 2 3

[1]. 13 9 11 5 12 7 10 3 8 y 1 x 14 2 6 4

[2]. 13 9 11 2 6 4 10 3 8 5 12 7 x 14 y 1.

Using the additive rule and not changing the symbols x and y we get the scheme:

0.	1	x	8	5	9	4	13	12	14	y	7	10	6	11	2	3
3.	4	x	11	8	12	7	1	0	2	y	10	13	9	14	5	6
6.	7	x	14	11	0	10	4	3	5	y	13	1	12	2	8	9
9.	10	x	2	14	3	13	7	6	8	y	1	4	0	5	11	12
12.	13	x	5	2	6	1	10	9	11	y	4	7	3	8	14	0
1.	14	10	12	6	13	8	11	4	9	y	2	x	0	3	7	5
4.	2	13	0	9	1	11	14	7	12	y	5	x	3	6	10	8
7.	5	1	3	12	4	14	2	10	0	y	8	x	6	9	13	11
10.	8	4	6	0	7	2	5	13	3	y	11	x	9	12	1	14
13.	11	7	9	3	10	5	8	1	6	y	14	x	12	0	4	2
2.	0	11	13	4	8	6	12	5	10	7	14	9	x	1	y	3
5.	3	14	1	7	11	9	0	8	13	10	2	12	x	4	y	6
8.	6	2	4	10	14	12	3	11	1	13	5	0	x	7	y	9
11.	9	5	7	13	2	0	6	14	4	1	8	3	x	10	y	12
14.	12	8	10	1	5	3	9	2	7	4	11	6	x	13	y	0
x.	0	1	2	9	10	11	3	4	5	12	13	14	6	7	8	
y.	3	2	1	9	8	7	0	14	13	6	5	4	12	11	10.	

As usual we manufactured row x and row y by applying Rule R^* (or Δ^*). The scheme describes an orientable triangular embedding of $K_{17}-K_2$.

Pick every third element in row x and you get the sequence 0, 9, 3, 12, 6 or 1, 10, 4, 13, 7 or 2, 11, 5, 14, 8. These are arithmetic sequences with difference $d=9$. On the other hand the sum of the three inward flowing currents at the vortex x in Fig. 9.10 is $d=9$. (The same situation is found at vortex y and row y with $d=6$.) This is not just an accident. One can easily prove the following general statement: *If at a vortex x in an index 3 solution, the sum of all three (counted as inward flowing) currents is d, then the row x manufactured by rule Δ^* has the form*

$$x. \quad a, b, c, a+d, b+d, c+d, a+2d, \ldots.$$

The proof is just as in Section 7.1 and we leave it to the reader. Since row x has the above mentioned form we now see why in property E 5) the sum d is supposed to be a generator of the subgroup of all multiples of 3. In fact property E 5) guarantees the existence of a suitable row x.

Fig. 9.11 illustrates a current graph with properties E1) to E6), which gives a triangular embedding of $K_{29}-K_2$. Notice that the distribution of the elements of the group $\mathbb{Z}_{27}$ is almost the same as in Fig. 9.8. The lower horizontals with the attached verticals again form an arithmetic comb. In fact it is precisely the same comb as in Fig. 9.8. We can use this remark to describe the general solution: Take the unfinished picture of Fig. 9.9 and change the current of the first vertical to the left from $6s+1$ to 1. Moreover change the rotation on both ends of this vertical and call the vertices x and y.

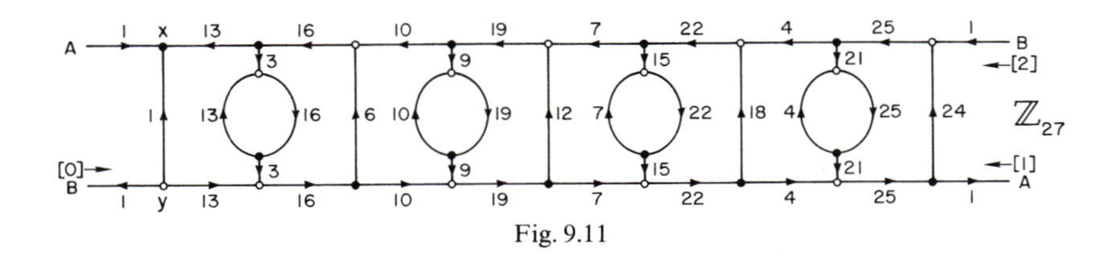

Fig. 9.11

Complete the rest in the obvious way such that the rungs with the currents 3, 9, 15, ... become globular rungs. Notice that the currents flowing into the vortex x are 1, 1 and $6s+1$. Their sum is $6s+3$ and the greatest common divisor $(6s+3,\ 12s+3)=(6s+3,\ -3)=3$. This takes care of property E5) for the vortex x. For the vortex y the sum of all the inward flowing currents is $-1-1-(6s+1)=-(6s+3)$ which also generates the subgroup of all multiples of 3. So we have constructed a triangular embedding of K_n-K_2 into an orientable surface for all $n\equiv 5$ (mod 12). It is obvious that we can perform the one missing adjacency with one additional handle.

From Theorem 5.7 it follows that the Eq. (4.13) about the genus of K_n is true for all $n\equiv 5$ (mod 12).

9.3. Orientable Case 6

Unfortunately this case is a little more complicated to explain. Let n be of the form $n=12s+6$. We will use the elements of the group $\mathbb{Z}_{12s+3}$ and the symbols x, y, z to identify the $n=12s+6$ vertices of the graph K_n-K_3. We will be able to show that a triangular embedding of K_n-K_3 into an orientable surface exists for all $s\geqq 2$. For $s=1$ see Section 5.2. The general solution will work only for $s\geqq 4$. Therefore we need an ad hoc solution for $s=2$ and $s=3$. The solution for $s=2$ is given by Fig. 9.12. The reader should check that properties E2) to E6) are satisfied. Obviously property E1) has to be replaced by

E 1*) *Each vertex is of valence* 3 *or* 1. Moreover in Fig. 9.12 we see three dead-end-arcs; all three carry the current 9. This means that the following property holds:

E 7) *An element of order* 3 *is allowed to be the current of a dead-end-arc* (no letter has to be at the vertex of valence 1).

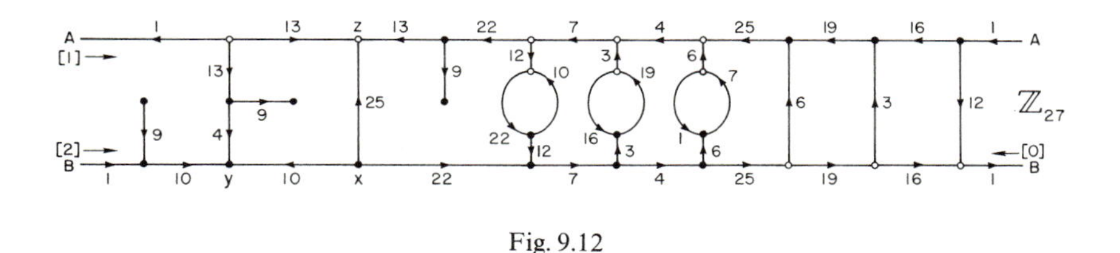

Fig. 9.12

This construction principle is familiar from the corresponding property C 8) for index 1 solutions. We will use E 7) only in the present ad hoc example $n=30$. Therefore we do not give a proof that E 7) works in general.

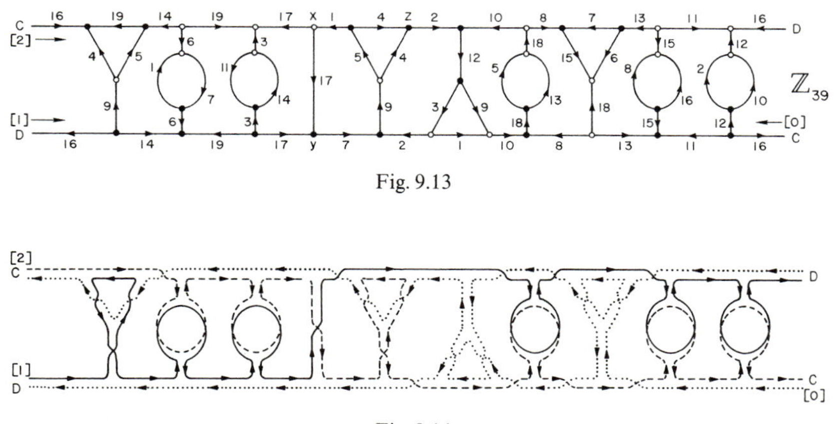

Fig. 9.13

Fig. 9.14

The Figs. 9.13 and 9.14 show the existence of an orientable triangular embedding of $K_{42}-K_3$ and Fig. 9.15 describes the general index 3 solution of Case 6 for all $s \geqq 4$. One can check that the index 3 conditions E 1) to E 6) are satisfied. Sorry, it takes time to do this. But at present Fig. 9.15 is the simplest known solution for Case 6 and was found by Guy and Youngs [31].

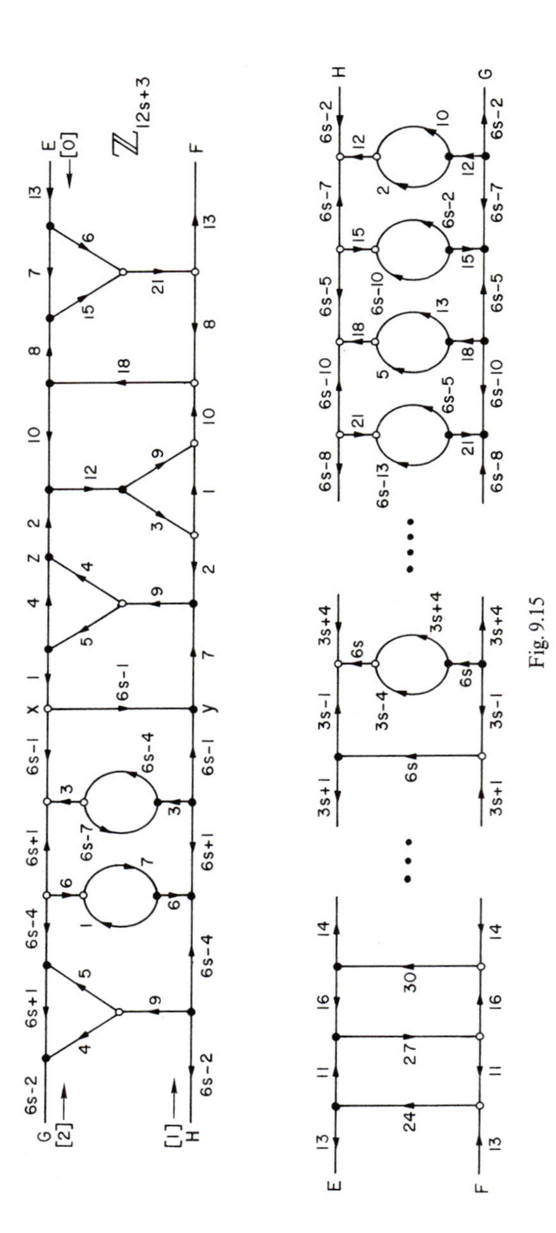

Fig. 9.15

The case $s=3$ illustrated in Figs. 9.13 and 9.14 is almost the same as the general case in Fig. 9.15. In fact it is easy to make Fig. 9.15 a little more complicated so that it works for all $s \geqq 3$.

9.4. Non-Orientable Case 9

This Case we will solve with an index 3 cascade. We use the elements of the group $\mathbb{Z}_{12s+9}$ to represent the vertices of K_{12s+9}. The cascade has three parts as in Fig. 9.16 ($s=1$). The two parts at the right side are exact duplicates of the cascade of Fig. 8.21 and the left part is obtained by multiplying the currents therein by -2. Since -2 is relatively prime to $12s+9$ this multiplication does not disturb the pertinent properties of cascades:

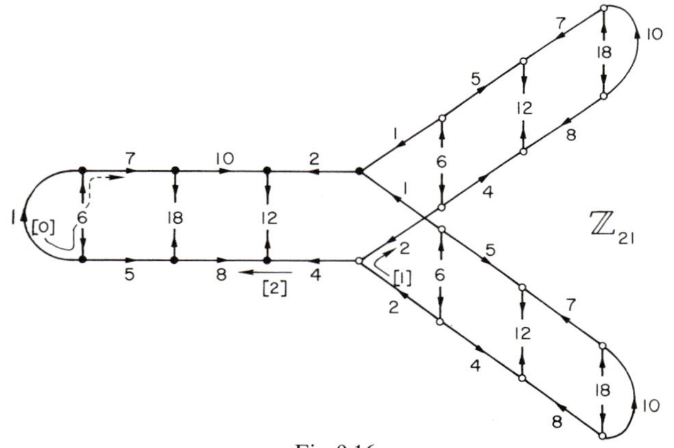

Fig. 9.16

In fact, if g is a current on an oriented arc in Fig. 8.21, then $-2g$, the new current on the oriented arc, is congruent to g (mod 3). Hence the current distribution, modulo 3, is the same in all three parts of the cascade. This remark makes it easy to check that the cascade of Fig. 9.16 ($s=1$) and its generalization for an arbitrary s satisfy the properties E1) to E5) of an index 3 solution (see Section 9.1). Property E5) does not apply and properties E2), E3) have to be understood in the sense of cascades. (Multiply the behavior by -1 at each midpoint of a broken arc.)

Property E6) must be generalized in the following manner.

E6*) *If along the arc with current c, either Fig. 9.7 holds* (i.e., *both circuits have normal behavior*), *or Fig. 9.17 holds* (*the behavior of circuit*

[a] *has been multiplied by* -1), *or Fig.* 9.18 *holds* (*the behavior of both circuits has been multiplied by* -1), *then* $c \equiv b-a$ (mod 3).

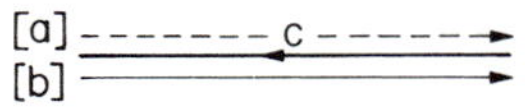

[a] - - - - - - - c - - - - - - →
[b] ——————————————→

Fig. 9.17

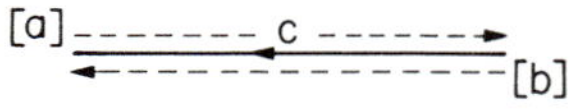

[a] - - - - - - - c - - - - - - →
← - - - - - - - - - - - - - - - [b]

Fig. 9.18

Property E6*) insures that rule R will be satisfied.

Notice that E6*) also makes sense if the arc in question is a broken one. When checking E6) it helps that all broken arcs in the cascade have currents $\equiv 0$ (mod 3).

The cascade produces a triangular embedding of K_{12s+9} for all $s \geqq 0$. The surface must be non-orientable because its Euler characteristic is an odd number.

Remark. The above method also works for the non-orientable Case 3. But there is a much easier solution available in this Case which is presented in Section 8.2.

10. Construction by Induction

10.1. An Index 3 Induction

With the help of the following theorem we will solve the non-orientable Case 1. We do not really need the orientable version of Theorem 10.1 because we already have easy solutions in Cases 1 and 7.

Theorem 10.1. *If there exists an (non-) orientable triangular embedding of* K_{2t+1} *for* $t \geqq 2$, *then there exists an (non-)orientable triangular embedding of* K_{6t+1}.

The following construction gives the proof.

Consider the group $\mathbb{Z}_{6t}$ and the current graph of Fig. 10.1 if $t=3$ and Fig. 10.2 for arbitrary $t \geqq 2$. It is of index 3 but does not satisfy all the properties of the list E 1) to E 6) from Section 9.1. In fact only E 2), E 4), and E 6) are true. The others hold in the following modified form.

E 1*) *Each vertex is of valence* 3 *or* 1.

E 3*) *Each element of the group* $\mathbb{Z}_{6t}$ *except* $0, \hat{3}, 6, 9, \ldots, 6t-6$ *appears exactly once in the log of each of the three circuits.*

E 5*) *For the vertex of valence* 3 *identified by a letter property* E 5) *is true.*

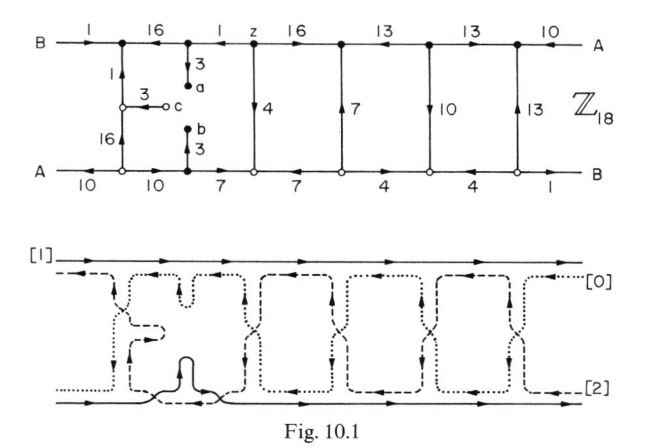

Fig. 10.1

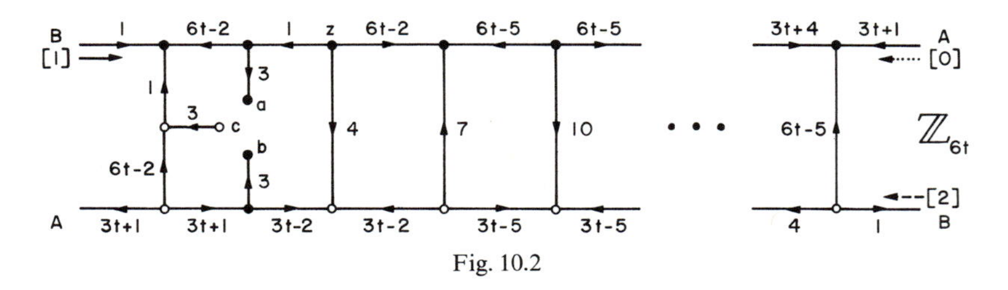

Fig. 10.2

E 7*) *Each of the three vertices a, b, c is of valence* 1 *and is incident with an end arc carrying current* 3 *respectively* -3. *Each of the three circuits* [0], [1], [2] *passes exactly one of the three vertices a, b, c.*

In the special case $t=3$ the logs of the three circuits given by the current graph of Fig. 10.1 read

[0]. 5 4 8 13 11 7 14 *z* 1 3 *a* 15 16 17 2 10
[1]. 1 2 17 *z* 16 5 13 8 10 3 *b* 15 7 11 4 14
[2]. 10 14 7 2 *z* 4 11 8 16 15 *c* 3 1 17 13 5.

Therefore the total scheme is as follows:

0. 5 4 8 13 11 7 14 *z* 1 3 *a* 15 16 17 2 10
3. 8 7 11 16 14 10 17 *z* 4 6 *a* 0 1 2 5 13
6. 11 10 14 1 17 13 2 *z* 7 9 *a* 3 4 5 8 16
9. 14 13 17 4 2 16 5 *z* 10 12 *a* 6 7 8 11 1
12. 17 16 2 7 5 1 8 *z* 13 15 *a* 9 10 11 14 4
15. 2 1 5 10 8 4 11 *z* 16 0 *a* 12 13 14 17 7

1. 2 3 0 *z* 17 6 14 9 11 4 *b* 16 8 12 5 15
4. 5 6 3 *z* 2 9 17 12 14 7 *b* 1 11 15 8 0
7. 8 9 6 *z* 5 12 2 15 17 10 *b* 4 14 0 11 3
10. 11 12 9 *z* 8 15 5 0 2 13 *b* 7 17 3 14 6
13. 14 15 12 *z* 11 0 8 3 5 16 *b* 10 2 6 17 9
16. 17 0 15 *z* 14 3 11 6 8 1 *b* 13 5 9 2 12

2. 12 16 9 4 *z* 6 13 10 0 17 *c* 5 3 1 15 7
5. 15 1 12 7 *z* 9 16 13 3 2 *c* 8 6 4 0 10
8. 0 4 15 10 *z* 12 1 16 6 5 *c* 11 9 7 3 13
11. 3 7 0 13 *z* 15 4 1 9 8 *c* 14 12 10 6 16
14. 6 10 3 16 *z* 0 7 4 12 11 *c* 17 15 13 9 1
17. 9 13 6 1 *z* 3 10 7 15 14 *c* 2 0 16 12 4

z. 14 16 15 11 13 12 8 10 9 5 7 6 2 4 3 17 1 0
a. 0 3 6 9 12 15
b. 1 4 7 10 13 16
c. 17 14 11 8 5 2.

The general scheme generated by the current graph of Fig. 10.2 describes a triangular embedding of a graph into an orientable surface. The graph is of course different from a complete graph. We consider the dual map M of this embedding. The map has $6t$ numbered countries, along with the countries z, a, b, and c. Notice that any two numbered countries i, j with $i \not\equiv j \pmod 3$ are adjacent and they are not adjacent if $i \equiv j \pmod 3$ unless they are in juxtaposition around country a, b or c. The country z is adjacent to all the numbered countries. We will now obtain the missing adjacencies between the numbered countries, and at the same time get rid of countries a, b and c from the map. This will give a map with $6t+1$ mutually adjacent countries,

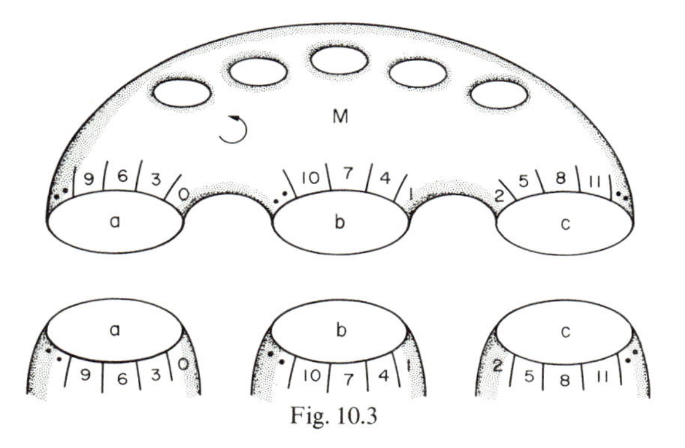

Fig. 10.3

Assume that there exists an orientable triangular embedding of K_{2t+1}. We consider the dual map, where all $2t+1$ countries are mutually adjacent. We take three copies of this map. (See the lower part of Fig. 10.3.) The countries of the first (second; third) copy we relabel as $a, 0, 3, 6, \ldots, 6t-3$ (as $b, 1, 4, 7, \ldots, 6t-2$; as $c, 2, 5, 8, \ldots, 6t-1$). We can do this in such a way that the cyclic order of the neighbors of country a (b; c) is

$$0, 3, 6, \ldots, 6t-3 \quad (1, 4, 7, \ldots, 6t-2; 2, 5, 8, \ldots, 6t-1).$$

Now we excise country a from M and from the first copy and identify the two boundaries in the obvious way (Fig. 10.3). We do the same with country b in M and the second copy and with country c in M and the third copy. We obtain a map on an orientable surface where all $6t+1$ countries are mutually adjacent and all vertices are of valence three. The dual is an orientable triangular embedding of K_{6t+1}. This proves the orientable part of Theorem 10.1.

Of course if the given embedding of K_{2t+1} is non-orientable instead of orientable then we obtain a non-orientable triangular embedding of K_{6t+1}. This completes the proof of Theorem 10.1.

There are two interesting modifications of the above construction.

We interchange the currents 1 and $6t-2$ of the two vertical arcs at the left of Fig. 10.2 and change the arrow of the arc incident with c. This creates two more vortices x and y as illustrated in Fig. 10.4. The vortices x and y satisfy, just as z does, the requirement of E 5) of Section 9.1. If we replace Fig. 10.2 by Fig. 10.4 in the above proof we get a proof for

Theorem 10.2. *If there exists an orientable triangular embedding of K_{2t+1} for $t \geqq 2$ then there exists an orientable triangular embedding of $K_{6t+3}-K_3$.*

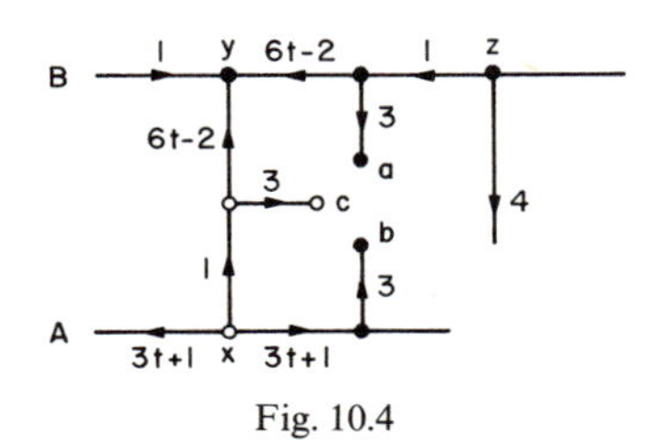

Fig. 10.4

If we apply this Theorem for the orientable Case 9 we only get two thirds of the solution, namely:

Theorem 10.3. *If* $s \not\equiv 2$ (mod 3) *and* $s \geqq 1$ *then there exists an orientable triangular embedding of* $K_{12s+9}-K_3$.

Proof. There exists an orientable triangular embedding of K_{4s+3} for all $s \equiv 1$ (mod 3), which is proved in Section 2.3 (Case 7), and for all $s \equiv 0$ (mod 3), which is proved in Section 9.2 (Case 3). Applying Theorem 10.2 with $t=2s+1$ gives Theorem 10.3.

The other modification of the construction which uses the current graph of Fig. 10.2 is a little more complicated.

Let t be of the form $t=6s+2$. Then a triangular embedding of K_{2t+1} does not exist. Therefore Theorem 10.1 is not applicable. But there exists an orientable triangular embedding of $K_{2t+1}-K_2$. We proved this in Section 9.2 (Case 5). Now use in the construction, as in Fig. 10.3, three copies of an embedding of $K_{2t+1}-K_2$ (instead of K_{2t+1}).

Obviously we obtain an orientable triangular embedding of a graph of the form $\{K_{6t+1}-3 \text{ arcs}\}$, because in each of the three copies there is one arc missing. We would like to obtain these three missing adjacencies by using one extra cross cap. It will be easy because we can judiciously select which of the three adjacencies are missing.

Take a glance at the scheme for $K_{17}-K_2$ or $K_{12s+5}-K_2$ produced by Figs. 9.10, 9.11 or their generalization described in Section 9.2; we can see that the rows 0 and 1 have the form

$$\begin{array}{l} 0. \ \ldots x \ldots y \ldots \\ 1. \ \ldots .y.x \ldots . \end{array}$$

In row 1 there is only one number between y and x. Row 0 tells us that x and y are exactly opposite countries in the cyclic oder of all neighbors around country 0. That means the scheme of the first copy of the dual of the triangular embedding of $K_{12s+5}-K_2=K_{2t+1}-K_2$ can be written so that row a reads

$$a. \ 0, 3, 6, \ldots, 6t-3,$$

and the two countries 0 and $6t-6$ are not adjacent. We write the second copy in such a way that row

$$b. \ 1, 4, 7, \ldots, 6t-2,$$

occurs and 7 and $3t+7$ are not adjacent. And let

$$c. \ 6t-1, \ldots, 8, 5, 2$$

be row c in the third copy and the two countries 14 and 8 are not adjacent.

Therefore the construction corresponding to Fig. 10.3 gives us a map with $6t+1$ countries where the three adjacencies

$$(0, 6t-6), \ (7, 3t+7), \ (8, 14)$$

are missing.

Now assume $t \geqq 8$.

In Fig. 10.2 the fourth vertical reading from the right to the left carries the current $6t-14$ and is traversed by the circuit [0] in a direction opposite its orientation. The arcs before and after this one have currents 7 and $3t+7$. Therefore

$$[0]. \ \ldots, 7, 14, 3t+7, \ldots$$

is part of circuit [0]. From Fig. 10.2 we also obtain the part

$$[1]. \ \ldots, 10, -10, 7, -7, 4, -4, \ldots$$

from circuit [1]. Therefore

$$\begin{array}{l} 0. \ \ldots, 7, 14, 3t+7, \ldots \\ 4. \ \ldots, 14, -6, \ldots, 8, 0, \ldots \end{array}$$

is part of the scheme and the partial picture of Fig. 10.5 occurs.

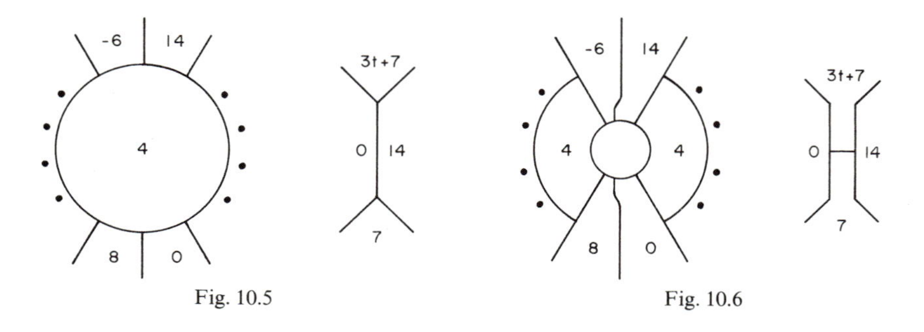

Fig. 10.5

Fig. 10.6

We transform Fig. 10.5 into Fig. 10.6 in the following way. Excise a circle out of the interior of country 4 and identify opposite pairs of points of the boundary. Then 4 becomes a cross cap or a Möbius strip. We extend the countries -6, 14, 8, and 0 using some territory of 4 as illustrated in Fig. 10.6. After this, country 4 is still connected. At the right of Fig. 10.6 we just change the adjacency (0, 14) into the adjacency $(7, 3t+7)$. This way we obtain a map with $6t+1$ mutually adjacent countries and only vertices of valence three on a non-orientable surface for every $t \equiv 2 \pmod 6$ and $t \geqq 8$. So we have proven for $s \geqq 4$:

Theorem 10.4. *If* $s \equiv 1 \pmod 3$ *then there exists a non-orientable triangular embedding of* K_{12s+1}.

From Section 8.3 (Fig. 8.22), we know Theorem 10.4 also holds for $s=1$.

10.2. An Index 2 Induction

This time we will use a very easy index 2 cascade like the index 1 cascades in Section 8.3. Consider the group $\mathbb{Z}_{4t+2}$ and the examples given in Figs. 10.7 ($t=2$) and 10.8 ($t=5$). The generalization of the cascade is so easy that we leave it to the reader. Just take t vertical broken arcs

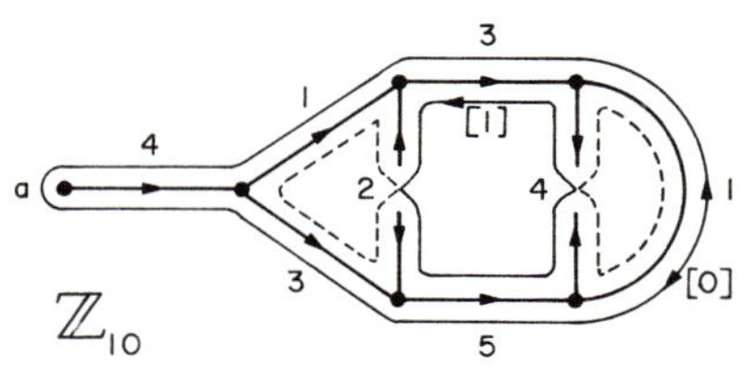

Fig. 10.7

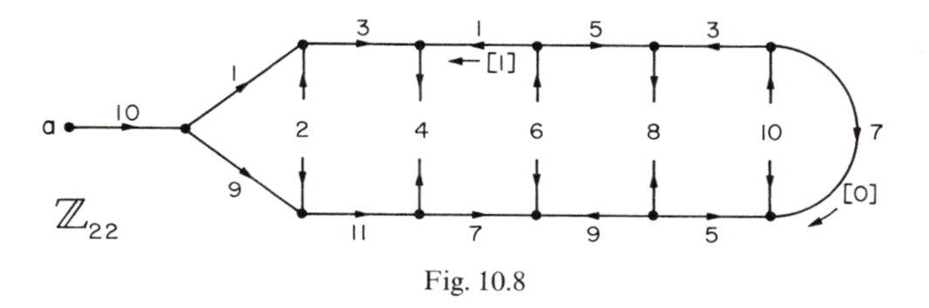

Fig. 10.8

carrying the currents 2, 4, 6, ..., $2t$. Their directions alternate. The horizontal dead-end-arc carries the current $2t$ and points to the right. The two nonhorizontal arcs of the triangle have currents 1 and $2t-1$ and point to the right. The rest is uniquely determined by KCL.

The cascade has the properties: KCL holds at each vertex of valence three. The current $2t$ on the dead-end-arc incident with vertex a generates the subgroup of all even elements of $\mathbb{Z}_{4t+2}$. The cascade induces two circuits [0] and [1]. (Remember that at the midpoint of each broken line one has to multiply the behavior by -1.) The log of circuit [1] contains each nonzero element of the group $\mathbb{Z}_{4t+2}$ exactly once and does not contain the symbol a. The log of circuit [0] contains all the odd elements and only two even ones, namely $2t$ and $-2t$. Between the two even elements there is the symbol a.

Each arc which occurs in both circuits carries an odd current. Otherwise the current is even. This agrees with the index 2 property D7) from Section 9.1.

Remark. In the cascade the lower horizontal on the left carries the element $2t+1$, which is of order 2. This is allowed in an index 2 solution.

As an example see the scheme

	0.	6	a	4	1	3	9	5	7	
	2.	8	a	6	3	5	1	7	9	
	4.	0	a	8	5	7	3	9	1	
	6.	2	a	0	7	9	5	1	3	
	8.	4	a	2	9	1	7	3	5	
(M)	1.	0	3	6	5	2	7	8	9	4
	3.	2	5	8	7	4	9	0	1	6
	5.	4	7	0	9	6	1	2	3	8
	7.	6	9	2	1	8	3	4	5	0
	9.	8	1	4	3	0	5	6	7	2
	a.	2	8	4	0	6				

generated by the cascade of Fig. 10.7 when $t=2$. It satisfies Rule R and defines a triangular embedding into a closed surface.

We consider the dual map M of this embedding for an arbitrary t. (Even $t=1$ is possible.) The country a is adjacent to all the even-numbered countries in a certain cyclic order. Two even-numbered countries are not adjacent to each other unless they are next to each other in row a.

Now assume there exists a non-orientable triangular embedding of K_{2t+2}. Consider the dual map N and use the elements a, 0, 2, 4, ..., $4t$ to identify the countries in such a way that row a is exactly the same permutation of $0, 2, 4, \ldots, 4t$ as in the row a from the scheme of the map M.

For instance for $t=2$ we write the scheme (5.3) for K_6 in the form

$$(N)\qquad \begin{array}{llllll} a. & 2 & 8 & 4 & 0 & 6 \\ 0. & 8 & 6 & a & 4 & 2 \\ 2. & 0 & 8 & a & 6 & 4 \\ 4. & 2 & 0 & a & 8 & 6 \\ 6. & 4 & 2 & a & 0 & 8 \\ 8. & 6 & 4 & a & 2 & 0. \end{array}$$

Excise country a out of M and country a out of N and identify the boundaries in the same way we already did in Fig. 10.3. We obtain a map with $4t+2$ mutually adjacent countries. This proves

Theorem 10.5. *If there exists a non-orientable triangular embedding of K_{2t+2} then there exists a non-orientable triangular embedding of K_{4t+2}.*

If a triangular embedding of K_{2t+2} does not exist but there is one of $K_{2t+2}-K_2$, then of course we can use the construction and we obtain a proof for

Theorem 10.6. *If there exists a non-orientable triangular embedding of $K_{2t+2}-K_2$ then there exists a non-orientable triangular embedding of $K_{4t+2}-K_2$.*

Since we do not have a proof that a non-orientable embedding of K_n-K_2 exists for $n\equiv 8 \pmod{12}$, this Theorem 10.6 is not very helpful for solving Case 2. But the following slight modification will do.

In the two non-regular non-orientable Cases 5 and 11 we found (Sec. 8.3) a triangular embedding of K_n-K_2, and, as shown in Figs. 5.3 and 5.4, the missing adjacency was made with one cross cap. The resulting dual map contained two extra adjacencies, giving a triangular embedding of K_n plus two additional arcs. The natural approach would be to do this for Cases 2 and 8 also, but in Case 8 we were unable to find a triangular embedding of K_n-K_2. We wish to use the solution of Case 8 for solving Case 2. In Case 8 we constructed a triangular embedding of K_n plus two additional arcs. Therefore we prove the following:

Theorem 10.7. *If there exists a non-orientable triangular embedding of K_{2t+2} plus two additional arcs, then there exists a non-orientable triangular embedding of K_{4t+2} plus two additional arcs.*

Proof. The hypothesis can only apply when $t \geqq 3$, as a simple application of Theorem 4.4 shows. Now we can proceed as in the proof of Theorem 10.5, identifying the countries of the dual map of the given triangular embedding so that country a is adjacent exactly once to each of the $2t+1$ remaining countries. This can be done since at most 4 countries have multiple adjacencies resulting from the extra arcs, and $t \geqq 3$. Then the 1-skeleton of the dual to the non-orientable triangular embedding resulting from the identification of the two surfaces in Theorem 10.5 is K_{4t+2} plus two additional arcs.

10.3. Non-Orientable Cases 1, 2, 6, and 10

Case 1. Now we can solve the rest of the non-orientable Cases by induction.

Theorem 10.8. *There exists a non-orientable triangular embedding of K_{12s+1} for each integer $s \geqq 1$.*

For the proof we consider three cases:

a) If $s \equiv 1 \pmod 3$ then Theorem 10.8 follows from Theorem 10.4.

b) If $s \equiv 2 \pmod 3$ then let s be of the form $s = 3t+2$. Since we solved Case 9 in Section 9.4, there exists a non-orientable triangular embedding of $K_{12t+9} = K_{4s+1}$. Using Theorem 10.1 we obtain the desired embedding of K_{12s+1}.

c) For this third case we use induction. Let $s = 3t$ and assume a triangular non-orientable embedding of $K_{12s'+1}$ exists for each $s' < s$.

Therefore a triangular non-orientable embedding for $K_{12t+1} = K_{4s+1}$ exists. By Theorem 10.1 the same holds for K_{12s+1}. This completes the proof of Theorem 10.8 (Case 1).

Cases 6 and 10. These two cases can be solved together.

Theorem 10.9. *There exists a non-orientable triangular embedding of K_{12s+6} and K_{12s+10} for every nonnegative integer s.*

Proof. From Section 5.1 we know that there exists a non-orientable triangular embedding of K_6 and by Theorem 10.5 ($t=2$) there also exists one of K_{10}. Therefore Theorem 10.9 holds for $s=0$.

Given $s>0$ and assume Theorem 10.9 is already proven for all smaller values of s.

There exists a non-orientable triangular embedding of K_{6s+4}. This follows from the already solved Case 4 (Section 8.3) if s is even and by the induction hypothesis if s is odd. Using Theorem 10.5 with $t=3s+1$ we obtain the desired embedding of K_{12s+6}.

There exists a non-orientable triangular embedding of K_{6s+6}. This follows from the already solved Case 0 (Section 8.2) if s is odd and by the induction hypothesis if s is even. Using Theorem 10.5 with $t=3s+2$ we obtain the desired embedding of K_{12s+10}. This completes the proof of Theorem 10.9.

Case 2. This Case can now be solved by induction.

Theorem 10.10. *There exists a non-orientable triangular embedding of K_{12s+2} plus two additional arcs for all $s\geqq 1$.*

Proof. In Section 8.3 we solved the non-orientable Case 8 for $s\geqq 0$. Examination of Fig. 8.29 will reveal that the non-orientable embedding produced by the current graph and modifications of Figs. 8.27, 8.28, and 8.29, is in fact a map with $12s+8$ countries, and only vertices of valence three, in which the adjacencies (α, y) and (β, y) occur twice, and all the other adjacencies occur only once. Therefore the 1-skeleton of the map dual to this triangular embedding is K_{12s+8} plus two additional arcs.

Letting $s=0$, we can apply Theorem 10.7 with $t=3$ to obtain a non-orientable triangular embedding of K_{14} plus two additional arcs. This is the beginning of the induction. Suppose $s>1$ and for all $s'<s$ there exists a non-orientable triangular embedding of $K_{12s'+2}$ plus two additional arcs. Then there exists a non-orientable triangular embedding of K_{6s+2} plus two additional arcs by this hypothesis if s is even and by the already solved Case 8 if s is odd. Applying Theorem 10.7 gives a non-orientable triangular embedding of K_{12s+2} plus two additional arcs.

Computing the genus of this embedding gives the right side of formula (4.19), proving it for all $n\equiv 2 \pmod{12}$.

This completes the determination of the non-orientable genus of the complete graph K_n for every n. For the orientable genus there is one Case left. We will do this in the next chapter.

11. Orientable Case 0

11.1. Currents from Non-Abelian Groups

Since this is a regular case we shall construct an orientable triangular embedding of K_{12s} using a current graph of index 1. But this time the currents will come from a non-abelian group of order $12s$, and it is to be expected that some of the construction principles for a current graph have to be a little modified.

We now write the group operation as multiplication and let ε be the identity element of the group.

Exactly as in the previous constructions, the row ε will be the log of the circuit in a current graph. Of course this time we write a^{-1} instead of $-a$. Let

$$\varepsilon. \quad \alpha_1, \alpha_2, \ldots, \alpha_{n-1}$$

be the row ε. Then we get all the other rows of the scheme by multiplying all members of row ε by a certain element from the *left* side:

$$\alpha. \quad \alpha\alpha_1, \alpha\alpha_2, \ldots, \alpha\alpha_{n-1}.$$

When we do this for each element α of the group we obtain the whole scheme.

We will use the construction principles C1), C2), C3), C4), C6), and C8). All, except C4), can be translated easily into the language of a multiplicatively written non-abelian group. For instance C3) now reads

C3) *Each non-identity element or its inverse in the group appears exactly once as a current on some arc, but if an element α occurs, then its inverse does not occur unless $\alpha = \alpha^{-1}$.*

The construction principle C4) needs to be modified because KCL is ambiguous if the group is non-abelian. We generalize C4) as follows:

C4) *Let ρ, σ, τ be the currents (all viewed as) leaving the vertex P of valence 3 and let ρ, σ, τ be the cyclic order given by the rotation at P. Then $\rho\sigma\tau = \varepsilon$.*

If we have a current graph with the properties C1) to C4), C6), and C8), using a group with $12s$ elements, then it produces a triangular rotation of K_{12s}. The proof is given in Sections 2.3, 6.1, and 8.1.

The group we will use has many elements of order two. According to C6), they must appear on dead-end-arcs. In the figures we indicate this kind of dead-end-arc by omitting the dead-end-vertex. (One can interpret these arcs as loops.) In the log of the circuit the current of order 2 is recorded only once.

There will also be more than one element of order 3 appearing on a dead-end-arc according to C8). In Fig. 8.1 both kinds of dead-end-arcs are shown in one example.

11.2. Examples

We denote by A_4 the alternating group acting on four letters, that is, the group of all even permutations of the set $\{1, 2, 3, 4\}$. A_4 has 12 elements. We have to study the elements of A_4 in detail. We use the cyclic notation for permutations in A_4. Thus $(a, b, \ldots, h)$ is the permutation which permutes $a, b, \ldots, h$ cyclically in the given order. If σ, τ are permutations, $\sigma\tau$ denotes the permutation taking i into $\sigma(\tau(i))$.

The 12 elements of A_4 are

$$\begin{aligned}&(1),\ (1\,2)(3\,4),\ (1\,3)(2\,4),\ (1\,4)(2\,3),\\ &(1\,2\,3),\ (1\,3\,2),\ (1\,2\,4),\ (1\,4\,2),\\ &(1\,3\,4),\ (1\,4\,3),\ (2\,3\,4),\ (2\,4\,3).\end{aligned}$$

Two permutations λ and μ are called *conjugate* if and only if $\lambda = \sigma\mu\sigma^{-1}$ for some permutation σ.

Lemma 11.1. *If* $(a_1, a_2, \ldots, a_m)$ *is any cyclic permutation and* σ *is any permutation, then*

$$\sigma(a_1, a_2, \ldots, a_m)\sigma^{-1} = (\sigma(a_1), \sigma(a_2), \ldots, \sigma(a_m)).$$

Thus two permutations are conjugate only if their disjoint cycle decompositions contain the same number of cycles of each length.

Proof. It suffices to show that

$$\sigma(a_1, a_2, \ldots, a_m)\sigma^{-1}(\sigma(a_i)) = \begin{cases}\sigma(a_{i+1}) & \text{if } i<m\\ \sigma(a_1) & \text{if } i=m\end{cases}$$

and that

$$\sigma(a_1, a_2, \ldots, a_m)\sigma^{-1}(x) = x \quad \text{if } x \neq \sigma(a_i) \text{ for any } i.$$

But these are trivial calculations. The second assertion of Lemma 11.1 follows from the fact that $\sigma(\lambda\mu)\sigma^{-1} = (\sigma\lambda\sigma^{-1})(\sigma\mu\sigma^{-1})$. Hence if $\tau = c_1 \cdot c_2 \ldots c_m$ where the c_i are disjoint cycles, then

$$\sigma\tau\sigma^{-1} = (\sigma c_1\sigma^{-1})(\sigma c_2\sigma^{-1}) \ldots (\sigma c_m\sigma^{-1})$$

and c_i has the same length as $\sigma c_i \sigma^{-1}$. This completes the proof. For example the permutation (1 2)(3 4) is not conjugate to (1 2 3).

Consider the set of elements

$$N = \{(1), (1\,2)(3\,4), (1\,3)(2\,4), (1\,4)(2\,3)\}.$$

It is closed under composition (multiplication) and hence N is a subgroup of A_4. Since N contains all the permutations in A_4 with disjoint cyclic decompositions of the form $(a, b)(c, d)$ we see that $\sigma N \sigma^{-1} \subseteq N$ for all σ in A_4. This means N is a normal subgroup of A_4. Also it is easy to check that each element in N has order 2 or 1. Let $\alpha, \beta \in N$ be two elements of N, then $\alpha\beta = (\alpha\beta)\alpha^2 = \alpha\beta\alpha\beta^2\alpha = (\alpha\beta)^2\beta\alpha = \beta\alpha$. Therefore N is abelian. So by the structure theorem for abelian groups N is isomorphic to $\mathbb{Z}_4$ or $\mathbb{Z}_2 \times \mathbb{Z}_2$. Since N has no element of order 4 the latter occurs and $N \cong \mathbb{Z}_2 \times \mathbb{Z}_2$. Of course this fact could have been checked directly without using the structure theorem.

Let $\alpha = (1\,2\,3)$, and set $A = \{(1), \alpha, \alpha^2\}$. Then A is the cyclic subgroup of A_4 generated by α.

Lemma 11.2. *Every element of A_4 is uniquely a product of the form $n a$ where $a \in A$ and $n \in N$.*

Before we prove Lemma 11.2 it should be mentioned that it does not imply that A_4 is the direct product of the two subgroups because it is *not* always true that $(n_1 a_1)(n_2 a_2) = n_1 n_2 a_1 a_2$.

Proof. Two such products $n_1 a_1$ and $n_2 a_2$ are equal if and only if

$$a_1 a_2^{-1} = n_1^{-1} n_2 .$$

The left side lies in A, and the right side lies in N. Since A and N have only the element (1) in common it follows $a_1 a_2^{-1} = \varepsilon = n_1^{-1} n_2$, and therefore $a_1 = a_2$ and $n_1 = n_2$. This proves uniqueness, and shows also that there are exactly $|A| \cdot |N| = 12$ distinct products of the given form. Since $|A_4| = 12$ they must exhaust A_4. This proves Lemma 11.2.

Now we label the elements of A_4 in a convenient way: Define

$$\begin{aligned} &\Theta_1 = (1\,4)(2\,3), \\ (11.1) \qquad &\Theta_2 = (1\,3)(2\,4), \\ &\Theta_3 = (1\,2)(3\,4), \end{aligned}$$

$$\begin{aligned} &s_i = \Theta_i \Theta_{i+1} \quad (i = 1, 2, 3) \\ (11.2) \qquad &t_i = \Theta_i \alpha \quad (i = 1, 2, 3) \\ &t_4 = \alpha . \end{aligned}$$

In the above definition of s_3, Θ_4 really means Θ_1. By calculation, all three of the s_i are distinct. Then by Lemma 11.2, the seven elements in (11.2) are all distinct. The following are trivial calculations.

$$\begin{aligned} \alpha^{-1}\Theta_1\alpha &= (3\,4)(1\,2) = \Theta_3 \\ \alpha^{-1}\Theta_2\alpha &= (3\,2)(1\,4) = \Theta_1 \\ \alpha^{-1}\Theta_3\alpha &= (3\,1)(2\,4) = \Theta_2 . \end{aligned} \tag{11.3}$$

Now $t_i^{-1} = \alpha^{-1}\Theta_i^{-1} = \alpha^{-1}\Theta_i = (\alpha^{-1}\Theta_i\alpha)\alpha^{-1}$ if $1 \leqq i \leqq 3$. Hence from (11.3) we see that

$$\begin{aligned} t_1^{-1} &= \Theta_3\,\alpha^{-1} \\ t_2^{-1} &= \Theta_1\,\alpha^{-1} \\ t_3^{-1} &= \Theta_2\,\alpha^{-1} \\ t_4^{-1} &= \alpha^{-1} . \end{aligned} \tag{11.4}$$

Since $\alpha^{-1} \neq \alpha$, none of the elements in (11.4) appear in (11.2). Hence the collection

$$\begin{aligned} &s_i && (i = 1, 2, 3), \\ &t_i && (i = 1, 2, 3, 4), \\ &t_i^{-1} && (i = 1, 2, 3, 4), \end{aligned} \tag{11.5}$$

has exactly 11 elements of A_4. Therefore the elements (11.5) exhaust the non-identity elements of A_4 and are now ready to be used as currents in Fig. 11.1.

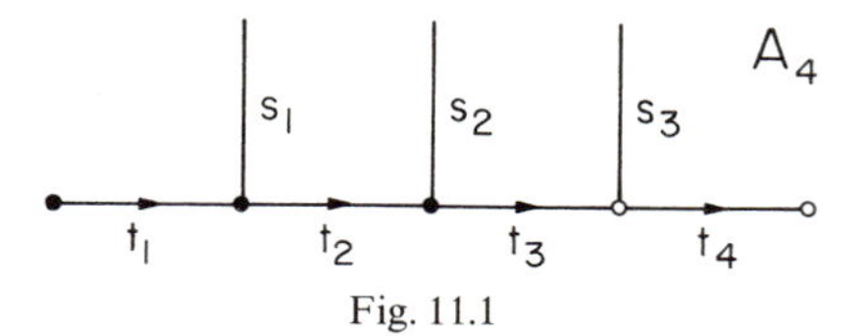

Fig. 11.1

We can easily check that construction principles C1) to C4), C6), and C8) all hold: Since $t_i^{-1} \neq t_i$ for each i, the three elements s_1, s_2, s_3 are all the elements of order 2. Hence C6) holds. By computation, t_1 and t_4 are of order 3, so that C8) holds. Direct calculation shows that $s_1 t_2 t_1^{-1} = s_2 t_3 t_2^{-1} = s_3 t_3^{-1} t_4 = \varepsilon$. Therefore KCL holds at every vertex of valence 3. The other construction principles are trivially verified. Thus the current graph of Fig. 11.1 produces an orientable triangular embedding of K_{12}.

Remark. The subgraph of Fig. 11.1 consisting only of the three arcs carrying the currents t_1, s_1, t_2 produces the icosahedron on the sphere or

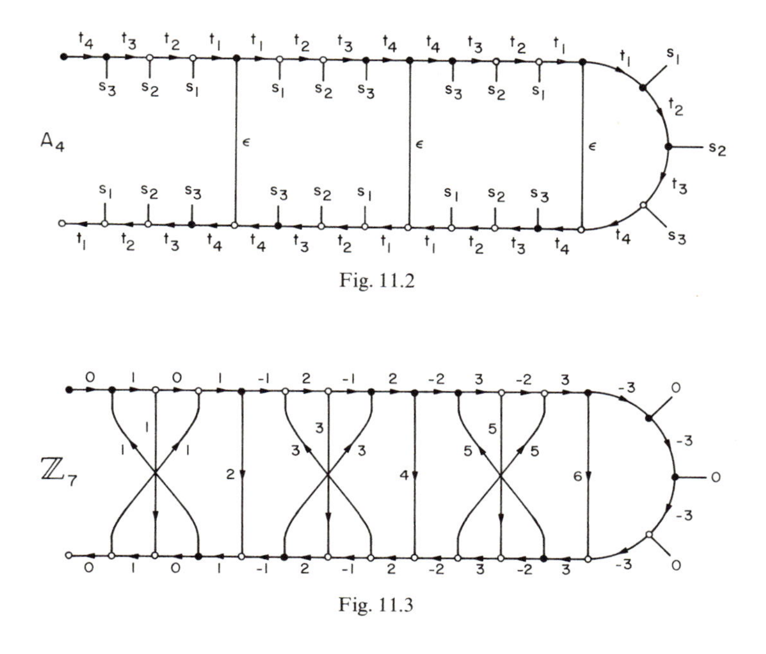

Fig. 11.2

Fig. 11.3

in dual form the dodecahedron. If, in Fig. 11.1, we reverse the orientation of all the currents and reverse the rotation at every vertex, it is easy to see KCL will still hold.

As a second example we exhibit a triangular orientable embedding of K_{84}. We will use the group $A_4 \times \mathbb{Z}_7$ which has $84 = 12 \cdot 7$ elements. Take 7 copies of the tree of Fig. 11.1 and arrange them as displayed in Fig. 11.2. There are 3 additional verticals carrying the identity ε of the group A_4. These three verticals divide the picture into three boxes. Then identify the arc s_i of each upper copy with the arc s_i of the lower copy in the same box as shown in Fig. 11.3. Then consider the graph of Fig. 11.3 as a current graph with currents (τ, i) from the group $A_4 \times \mathbb{Z}_7$, where $\tau \in A_4$ and $i \in \mathbb{Z}_7$. The distribution of the first components (elements of A_4) is displayed in Fig. 11.2. The second components (elements of $\mathbb{Z}_7$) are found in Fig. 11.3.

It is not difficult to check that the index 1 construction principles C1) to C4), C6), and C8) are all true:

The currents on the two horizontal dead-end-arcs at the left side of Fig. 11.3 are $(t_4, 0)$ and $(t_1, 0)$. Both are of order 3 and therefore C8) holds. There are three elements of order 2 in the group $A_4 \times \mathbb{Z}_7$. Each appears on a dead-end-arc at the right side of Fig. 11.3. This agrees with C6). The rotation chosen for the graph induces one single circuit. The rest of the properties can also be checked. Therefore there exists an orientable triangular embedding of K_{12s} if $s=7$.

There is an obvious way to generalize the above construction to obtain an embedding of K_{12s} if $s=2r+1$ is an odd number: We use the group $A_4\times\mathbb{Z}_s$. Then take s copies of the tree in Fig. 11.1 and arrange them as in Fig. 11.2 with r additional verticals and complete the graph as in Fig. 11.3. The second components of the currents on the horizontals are, reading from left to right:

$$0, 1, 0, 1, -1, 2, -1, 2, -2, 3, -2, 3, \ldots, 1-r, r, 1-r, r.$$

The verticals carry the elements

$$1, 1, 1, 2, 3, 3, 3, 4, 5, 5, 5, \ldots, s-2, s-2, s-2, s-1$$

in that order. The four curved arcs on the right all have the current $-r$. This current graph with currents from $A_4\times\mathbb{Z}_s$ generates a triangular orientable embedding of K_{12s} if s is odd.

11.3. General Solution

The group which will be used can be described as follows: Given the number $n=12s$, choose r and k so that $s=2^{k-2}(2r+1)$. Hence 2^{k-2} is the highest possible power of 2 which divides s. The group we will use is

$$A_4\times\underbrace{\mathbb{Z}_2\times\mathbb{Z}_2\times\cdots\times\mathbb{Z}_2}_{(k-2)\text{ times}}\times\mathbb{Z}_{2r+1}.$$

It has exactly $12\cdot 2^{k-2}\cdot(2r+1)=12s$ elements. Unfortunately a representation of the coming current graph using the group explicitly is not known.

We will use a very strange representation of this group. It seems necessary to consider the additive and multiplicative structures of finite fields. The following properties are well known to algebraists (see for instance Fraleigh [18], Lang [50], Van der Waerden [88]).

FF1) For every positive integer k there exists a finite field $GF(2^k)$ of order 2^k. (It is called the Galois field of order 2^k.)

The additive group in $GF(2^k)$ is designated as $F^+(2^k)$ and its identity element as 0. The elements $\neq 0$ form a group under multiplication, the multiplicative group $F^*(2^k)$.

FF2) The multiplicative group $F^*(2^k)$ is cyclic of order 2^k-1.

FF3) If Θ is a generator of $F^*(2^k)$, then $1, \Theta, \Theta^2, \ldots, \Theta^{k-1}$ is a *basis* for $GF(2^k)$ over $GF(2)$.

FF4) If $p\in F^+(2^k)$, then $p+p=0$.

Let Θ be any generator of $F^*(2^k)$. Using exponential notation for an automorphism, define an automorphism α of $F^+(2^k)$ by linear extension of the following mapping of the basis from FF 3):

$$(11.6)\qquad \begin{aligned} (\Theta^i)^\alpha &= \Theta^i \qquad (i=2,3,\ldots,k-1),\\ \Theta^\alpha &= 1+\Theta,\\ 1^\alpha &= \Theta. \end{aligned}$$

Notice that

$$1^{\alpha^3} = \Theta^{\alpha^2} = (1+\Theta)^\alpha = 1^\alpha + \Theta^\alpha = \Theta + 1 + \Theta = 1,$$
$$\Theta^{\alpha^3} = (1+\Theta)^{\alpha^2} = (\Theta+1+\Theta)^\alpha = 1^\alpha = \Theta.$$

This means α^3 is the identity automorphism e and the elements $\alpha, \alpha^2, \alpha^3 = e$ form a cyclic group A of order 3.

Now we define the group $G(k)$ which has order $3 \cdot 2^k$ and elements $[p, a]$, where $p \in F^+(2^k)$ and $a \in A$. Group multiplication is defined by

$$(11.7)\qquad [p,a]\cdot[q,b] = [p+q^a, ab].$$

It is easy to show that this operation is associative. The element $[0, e] = \varepsilon$ is the identity in $G(k)$. The inverse exists and is

$$(11.8)\qquad [p,a]^{-1} = [p^{a^{-1}}, a^{-1}].$$

This group $G(k)$ is sometimes denoted as the semidirect product $G(k) = F^+(2^k) * A$ of the two groups $F^+(2^k)$ and A.

Since we will use this group $G(k)$ in a current graph we are interested in knowing which elements are of order 2 and 3.

Lemma 11.3. *The element $[p, a]$ is of order 2 if and only if $p \neq 0$ and $a = e$.*

Proof. If $[p, a]$ is of order 2, then

$$[p,a]^2 = [p+p^a, a^2] = [0, e],$$

which implies that $a = e$, since A is cyclic of order 3. If $p = 0$, then $[p, a]$ is the identity and not of order 2. Hence $p \neq 0$ and $a = e$.

The converse statement follows from FF 4) immediately.

Lemma 11.4. *Both elements $[0, \alpha]$ and $[\Theta, \alpha]$ are of order 3.*

Proof. $[0,\alpha]^3 = [0, \alpha^3] = [0, e]$.

$$\begin{aligned}[\Theta,\alpha]^3 &= [\Theta+\Theta^\alpha, \alpha^2][\Theta,\alpha] = [\Theta+1+\Theta, \alpha^2][\Theta,\alpha]\\ &= [1,\alpha^2][\Theta,\alpha] = [1+\Theta^{\alpha^2}, \alpha^3] = [0, e].\end{aligned}$$

Let $b=2^k$ and define

$$(11.9)\qquad \begin{aligned} s_i &= [\Theta^i+\Theta^{i+1}, e] && (i=1,2,\ldots,b-1),\\ t_i &= [\Theta^i, \alpha] && (i=1,2,\ldots,b-1),\\ t_b &= [0, \alpha]. \end{aligned}$$

Applying Lemmas 11.3, 11.4, and FF3), observe that t_1 and t_b are of order 3 and each s_i has order 2.

Lemma 11.5. *The elements* s_i $(i=1,2,\ldots,b-1)$ *are all distinct.*

Proof. Let $1\leqq j<i\leqq b-1$ and $s_j=s_i$. It follows

$$\begin{aligned} \Theta^j+\Theta^{j+1} &= \Theta^i+\Theta^{i+1},\\ \Theta^j(1+\Theta) &= \Theta^i(1+\Theta),\\ \Theta^j &= \Theta^i, \end{aligned}$$

which is a contradiction because Θ generates the group $F^*(2^k)$.

By (11.8) the second component of t_i^{-1} is $\alpha^{-1}\neq\alpha$; hence the collection

$$(11.10)\qquad \begin{aligned} &s_i && (i=1,2,\ldots,b-1),\\ &t_i && (i=1,2,\ldots,b),\\ &t_i^{-1} && (i=1,2,\ldots,b), \end{aligned}$$

consists of $3b-1$ distinct elements and by FF3) none of them equals $[0,e]=\varepsilon$. Therefore the collection (11.10) exhausts all non-identity elements of the group $G(k)$.

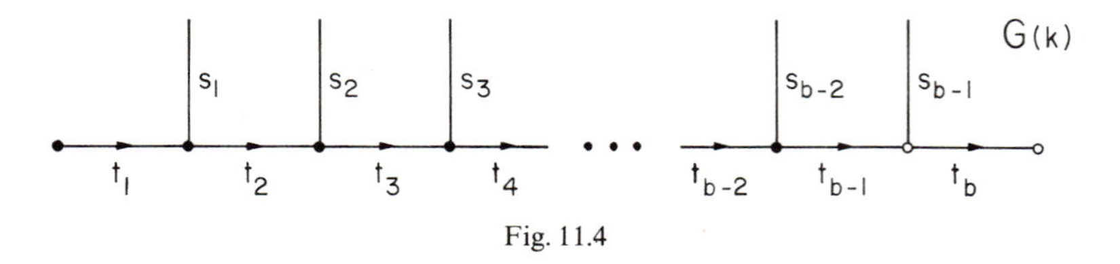

Fig. 11.4

Now we use the elements of the collection (11.10) as currents in the current graph shown in Fig. 11.4. From the properties of the elements s_i and t_i we already know it follows that construction principles C1) to C3), C6), and C8) are satisfied. We only have to show that KCL holds at each vertex of valence 3, i.e.,

$$t_i^{-1}\, s_i\, t_{i+1}=\varepsilon \quad \text{for } i=1,2,\ldots,b-2$$

and

$$t_b\, s_{b-1}\, t_{b-1}^{-1}=\varepsilon.$$

Proof.
$$s_i t_i = [\Theta^i + \Theta^{i+1}, e][\Theta^i, \alpha]$$
$$= [\Theta^i + \Theta^{i+1} + \Theta^i, \alpha] = t_{i+1}$$

holds for all $i \leqq b-2$. Therefore $t_i^{-1} s_i t_{i+1} = t_i^{-1} s_i s_i t_i = \varepsilon$ since each s_i is of order 2.

Since $b = 2^k$, by FF2):

$$\Theta^{b-1} = 1 \quad \text{and} \quad \Theta^b = \Theta.$$

Hence

$$t_{b-1} s_{b-1} = [1, \alpha][1 + \Theta, e] = [1 + 1^\alpha + \Theta^\alpha, \alpha]$$
$$= [1 + \Theta + 1 + \Theta, \alpha] = t_b,$$

so

$$t_b s_{b-1} t_{b-1}^{-1} = t_{b-1} s_{b-1} s_{b-1} t_{b-1}^{-1} = \varepsilon.$$

Since the current graph of Fig. 11.4 has all the required properties there exists a triangular orientable embedding of K_{12s} if s is a power of 2, namely $s = 2^{k-2}$.

Now we can easily explain the general solution for Case 0. Let $n \equiv 0 \pmod{12}$, then $n = 3 \cdot 2^k \cdot (2r+1)$. Take the group

$$G(k) \times \mathbb{Z}_{2r+1}.$$

The current graph we will use is built exactly as in Figs. 11.2 and 11.3. This time we just take $2r+1$ copies of the graph in Fig. 11.4. The desired properties of the current graph are easily verified.

We would like to mention that it is crucial that the number of teeth in the comb of Fig. 11.4 is odd. Otherwise the rotation would not induce one single circuit in the graph in question.

12. Related Problems

12.1. Questions about Rotations

Fig. 2.6 shows a rotation of K_5 which induces one single circuit. One may ask the more general question whether this is possible for K_n. Applying Theorem 2.5 one can see that each rotation of K_n induces an even number of circuits if $n \equiv 3$ or $0 \pmod 4$ and an odd number of circuits if $n \equiv 1$ or $2 \pmod 4$. Denote the vertices of K_n by $0, 1, 2, \ldots, n-1$ and consider the following rotation of K_n.

$$\begin{array}{rcccccccccccc}
0. & \ldots & & 5 & n-3 & 4 & n-2 & 3 & n-1 & & & 2 & 1 \\
1. & 2 & 3 & 4 & . & . & . & . & . & . & . & n-1 & 0 \\
2. & 3 & 4 & 5 & . & . & . & . & . & . & . & 0 & 1 \\
3. & 4 & 5 & 6 & . & . & . & . & . & . & . & 1 & 2 \\
\vdots & \vdots & \vdots & \vdots & \vdots & \vdots & \vdots & \vdots & \vdots & \vdots & \vdots & \vdots & \vdots \\
n-1. & 0 & 1 & 2 & . & . & . & . & . & . & . & n-3 & n-2.
\end{array}$$

One can check that this rotation of K_n induces one single circuit if $n \equiv 1$ or $2 \pmod 4$. If $n \equiv 3$ or $0 \pmod 4$ then the above rotation of K_n induces two circuits. This question was first solved by Nordhaus, Stewart, and White [57]. They used a more complicated rotation of K_n. The same question has also been solved for the bipartite graph $K_{m,n}$.

A graph consisting of only one vertex and having no arc is called *trivial*. The *connectivity* of a connected graph G is the smallest number of vertices whose removal results in either a disconnected graph or a trivial graph.

The graph G_4 of Fig. 2.1 has the property that each rotation induces more than two circuits (in fact more than 3). The connectivity of G_4 is 1.

To my knowledge, it is an unsolved question whether each graph with connectivity $\geqq 2$ has a rotation inducing only one or two circuits.

There are some interesting questions about infinite graphs. Let us consider only graphs with countably many vertices. A rotation at a vertex P of valence ∞ is a certain sequence $\ldots k_{-2}, k_{-1}, k_0, k_1, k_2 \ldots$ (extending infinitely in both directions) of all arcs k_i incident with P.

We also have to consider circuits extending infinitely in both directions. For instance each rotation of an infinite (in both directions) comb induces two infinite circuits. Or consider an infinite ladder-like graph (like Fig. 6.6, but both sides extend infinitely). If we give each upper vertex a clockwise rotation and each lower vertex a counter-clockwise rotation, then the rotation induces four infinite circuits.

The following seems to be an open question: Does there exist an infinite graph with a rotation where all vertices are of valence smaller than a given constant such that the rotation induces one single infinite (in both directions) circuit.

It is also not known whether there exists a triangular rotation of K_∞. The symbol ∞ refers to a countable set of vertices. In other words the problem is: Find a scheme

$$i.\ \ldots c_{-2}^{(i)}, c_{-1}^{(i)}, c_0^{(i)}, c_1^{(i)}, c_2^{(i)} \ldots \qquad (i = \ldots -2, -1, 0, 1, 2, \ldots)$$

such that $\ldots c_{-2}^{(i)}, c_{-1}^{(i)}, c_0^{(i)}, c_1^{(i)}, c_2^{(i)} \ldots$ is a permutation of all integers except i and the scheme satisfies Rule Δ^*. Another unsolved question is to find such a scheme for K_∞ satisfying only Rule Δ.

12.2. Questions about Embeddings

The following interesting conjecture is stated by R.H. Fox.

Conjecture. *Let G be a graph with exactly* 12 *vertices of valence* 5 *and all other vertices of valence* 6. *If the number of vertices of G is odd, then G is not planar.*

An embedding of a graph G into a closed surface S is called a *cellular* embedding if G divides S into only 2-cells (lunes, triangles, quadrilaterals, and so forth). If S is orientable the embedding induces in a very natural way a rotation of G. Since there are two different ways to orient S, the embedding induces two different rotations of G. One is exactly the opposite of the other.

The following theorem is a generalization of Theorem 5.6 and was first stated by Edmonds [15]. But Heffter [37] already knew it because he used it often, and not only for triangular embeddings.

Theorem of Heffter and Edmonds. *Given a graph with rotation, there exists a cellular embedding of the graph into an orientable surface such that one of the two rotations induced by the embedding is exactly the given rotation of the graph.*

Proof. Given a graph G with rotation. Identify each arc by a letter $a, b, c, \ldots$ and choose an orientation for each arc. Then write down the log of each circuit induced by the given rotation.

For instance, if G is the graph with the rotation shown in Fig. 2.4, we obtain the scheme

$$\begin{matrix} a & b & c & d & e & f \\ g & a^{-1} & c^{-1} & h & d^{-1} & f^{-1} \\ & b^{-1} & g^{-1} & e^{-1} & h^{-1} & \end{matrix}$$

if we label the arcs in an appropriate way. In any case the resulting scheme has the property that each letter a appears exactly twice, once in the inverse form a^{-1}. This means the scheme defines an orientable polyhedron. The 1-skeleton is the given graph and the orientation of the surface induces exactly the given rotation of the graph. Moreover the circuits induced by the rotation correspond in a natural way to the boundaries of the faces.

All the previous chapters were dedicated to determining the genus of the complete graph K_n. There are only a few more classes of graphs for which the genus is known. If the graph G has "enough" triangles then it can be expected that the genus of G equals the right side of (4.11). But there is not much hope for a general proof of such a theorem.

Let us consider the genus of the almost complete graph $K_n - K_2$. The first of the following sequence of inequalities comes from (4.11):

$$\left\{\frac{(n-3)(n-4)-2}{12}\right\} \leqq \gamma(K_n - K_2) \leqq \gamma(K_n) = \left\{\frac{(n-3)(n-4)}{12}\right\}.$$

The right and left side have the same value if $n \not\equiv 5 \pmod{12}$ and $n \not\equiv 2 \pmod{12}$.

A triangular embedding of $K_n - K_2$ is given in Section 9.2 for $n \equiv 5 \pmod{12}$, and by Ringel and Youngs [77] for all $n \equiv 14 \pmod{24}$. Recently Jungerman [41] solved the remaining case $n \equiv 2 \pmod{24}$ by a somewhat complicated index 2 solution. Therefore a triangular orientable embedding of $K_n - K_2$ exists if $n \equiv 2$ or $5 \pmod{12}$. Using Theorem 4.4 we obtain the general result

$$\gamma(K_n - K_2) = \left\{\frac{(n-3)(n-4)-2}{12}\right\}. \tag{12.1}$$

If we try to determine the genus of $K_n - K_3$ we run into some trouble. First it is not known whether there exist orientable triangular embeddings of the three graphs $K_9 - K_3$, $K_{13} - K_3$, and $K_{18} - K_3$.

Theorem 12.1. *Let $n \geqq 21$. Then the graph $K_n - K_3$ can be triangularly embedded into an orientable surface if and only if $n \equiv 1, 6, 9$, or $10 \pmod{12}$.*

Proof. If $K_n - K_3 \lhd S_p$, then Theorem 4.4 shows that $\alpha_1 = 3\alpha_0 - 3E(S_p)$. It follows

$$n(n-1) = 6n - 6(2-2p) + 6,$$

$$n(n-7) = 12(p-1) + 6.$$

Therefore $n(n-7)\equiv 6 \pmod{12}$ which means $n\equiv 1, 6, 9$, or $10 \pmod{12}$.

Conversely it is shown in this book that there exists a triangular orientable embedding of $K_n - K_3$ if $n\equiv 1, 6$ or $10 \pmod{12}$ and $n\geqq 21$. (See Sections 2.3, 6.3, 9.3.) But for $n\equiv 9 \pmod{12}$, J.W.T. Youngs [102] has given a proof ($n>9$). Compare this with Theorem 10.3, which gives only an incomplete answer.

As we saw previously for $K_n - K_2$, using (4.11) we can show that $\gamma(K_n - K_3)=\gamma(K_n)$ if $n\equiv 0, 3, 4$ or $7 \pmod{12}$. This leads, together with Theorem 12.1 and (12.1), to the result:

$$\gamma(K_n - K_3)=\left\{\frac{(n-3)(n-4)-6}{12}\right\} \tag{12.2}$$

for all $n\not\equiv 8$ and $11 \pmod{12}$ with $n\geqq 21$. For the two residue classes $n\equiv 8$ and $11 \pmod{12}$ it is not known whether (12.2) holds.

It is not very difficult to determine the genus of the complete bipartite graph $K_{m,n}$. For definition, see exercise 4.1.3.

Let $p=\gamma(K_{m,n})$. The shortest closed way in the graph $K_{m,n}$ has length 4. So we can use Theorem 4.5 with $k=4$:

$$2\alpha_1 \leqq 4\alpha_0 - 4(2-2p).$$

Since $\alpha_0 = m+n$ and $\alpha_1 = nm$ it follows that

$$\left\{\frac{(m-2)(n-2)}{4}\right\}\leqq \gamma(K_{m,n}). \tag{12.3}$$

Let m and n be even numbers. Then we can easily construct an orientable embedding of $K_{m,n}$ where all faces are quadrilaterals. We will show this in the case of $K_{6,4}$. Let 1, 2, 3, 4, 5, 6 be the vertices of the first class and a, b, c, d the four vertices of the second class.

Consider the rotation of $K_{6,4}$:

1. a b c d
2. d c b a
3. a b c d
4. d c b a
5. a b c d
6. d c b a

a. 1 2 3 4 5 6
b. 6 5 4 3 2 1
c. 1 2 3 4 5 6
d. 6 5 4 3 2 1.

It is easy to check that all the circuits induced by this rotation of $K_{6,4}$ have length 4. By the Theorem of Heffter and Edmonds there exists an orientable embedding of $K_{6,4}$ where all faces are quadrilaterals. The rotation can easily be generalized for every even m and n.

For a quadrilateral embedding of $K_{m,n}$ we have

$$4\alpha_2=2\alpha_1, \qquad \alpha_0=m+n, \qquad \alpha_1=mn,$$
$$\alpha_0-\alpha_1+\alpha_2=2-2p.$$

If we eliminate the α_i in this equations we obtain the result that (12.3) holds as an equality if m and n are even. For all the other cases it has also been shown by Ringel [70] that equality holds in (12.3).

For the tripartite graph $K_{n,n,n}$ the genus equals

$$\gamma(K_{n,n,n})=\frac{(n-1)(n-2)}{2}$$

for all $n\geqq 1$. See Ringel and Youngs [78]. A. White [89] generalized this formula for all graphs of the form $K_{an,n,n}$.

The formula

$$\gamma(K_{n,n,n,n})=(n-1)^2$$

has been proven recently for all $n\equiv 2 \pmod 4$ by B. Garman [23] and is conjectured for all n.

The following graph with 2^n vertices is called the n-dimensional cube Q_n. Its vertices are represented by all the sequences $(a_1, a_2, \dots, a_n)$ with $a_i=\pm 1$. Two vertices are adjacent if these sequences differ in exactly one place. The genus of Q_n was determined by Ringel [69] and later independently by Beinecke and Harary [4]:

$$\gamma(Q_n)=(n-4)\,2^{n-3}+1.$$

For determination of the genus of other classes of graphs see Alpert [1], Gross [25], Jacques [40], Himelwright [39], and White [90], [91], [92], [93].

It should be mentioned that the non-orientable genus of K_n-K_2 is not known for $n\equiv 8 \pmod{12}$. Although it agrees with the Euler formula a triangular embedding of K_8-K_2 does not exist (see Ringel [68]).

Let G be a graph. The maximum genus of an orientable surface S into which G is cellularly embeddable is denoted by $\gamma_M(G)$. It is called the maximum genus of G. Assume $p=\gamma_M(G)$. Therefore there exists a cellular embedding $G\subset S_p$. From Euler's formula $\alpha_0-\alpha_1+\alpha_2=2-2p$ and $\alpha_2\geqq 1$ we obtain

$$\gamma_M(G)\leqq\left[\frac{\alpha_1-\alpha_0+1}{2}\right]. \tag{12.4}$$

If there exists a cellular embedding of G into an orientable surface with $\alpha_2=1$ or 2, then (12.4) really holds as an equality.

For instance for the complete graph K_n we know that there exists a rotation inducing only one or two circuits (see the beginning of Section 12.1). Therefore by the Theorem of Heffter and Edmonds there exists an orientable cellular embedding of K_n with $\alpha_2=1$ or 2. This means

$$\gamma_M(K_n)=\left[\frac{(n-1)(n-2)}{4}\right] \quad \text{for } n\geqq 1.$$

For $K_{m,n}$ and Q_n the maximum genus is also known. R. D. Ringeisen [63] found that

$$\gamma_M(K_{m,n})=\left[\frac{(n-1)(m-1)}{2}\right] \quad \text{for } m,n\geqq 1$$

and J. Zaks [104] has shown that

$$\gamma_M(Q_n)=(n-2)\,2^{n-2} \quad \text{for } n\geqq 2.$$

We would like to mention two open problems about the crossing number $cr(G)$ of a graph G. This number $cr(G)$ is the smallest number of intersecting pairs of arcs in an "embedding" of G into the plane. For instance $cr(K_5)=1$, $cr(K_6)=3$. In general it is conjectured that

$$cr(K_n)=\frac{1}{4}\left[\frac{n}{2}\right]\left[\frac{n-1}{2}\right]\left[\frac{n-2}{2}\right]\left[\frac{n-3}{2}\right].$$

It is only known that the right side is an upper bound for $cr(K_n)$ (see Guy [30]).

For the bipartite graph there is a similar conjecture

$$cr(K_{m,n})=\left[\frac{m}{2}\right]\left[\frac{m-1}{2}\right]\left[\frac{n}{2}\right]\left[\frac{n-1}{2}\right]. \tag{12.5}$$

D. J. Kleitman [47] has proved (12.5) for all $m\leqq 6$ and any n.

References

1. Alpert, S. R.: The Genera of Amalgamations of Graphs. Ph. D. Thesis. Columbia University 1971.
2. Auslander, L., Brown, T. A., Youngs, J. W. T.: The imbedding of graphs in manifolds. J. Math. Mech. **12**, 629–634 (1963).
3. Battle, J., Harary, F., Kodama, Y., Youngs, J. W. T.: Additivity of the genus of a graph. Bull. Amer. Math. Soc. **68**, 565–568 (1962).
4. Beineke, L. W., Harary, F.: The genus of the n-cube. Canad. J. Math. **17**, 494–496 (1965).
5. Berge, C.: The Theory of Graphs and Its Applications. New York: John Wiley and Sons, Inc. 1962.
6. Bose, R. C.: On the construction of balanced incomplete block designs. Ann. of Eugenics **9**, 353–399 (1939).
7. Busacker, R. G., Saaty, T. L.: Finite Graphs and Networks: An Introduction with Applications. New York: McGraw-Hill Book Company 1965.
8. Coxeter, H. S. M.: The Map-Coloring of Unorientable Surfaces. Duke Math. J. **10**, 293–304 (1943).
9. Coxeter, H. S. M.: The Mathematics of Map Coloring. Journal of Recreational Mathematics **2**, No. 1, 1–14 (1969).
10. Dirac, G. A.: Map Colour Theorems. Canad. J. Math. **4**, 480–490 (1952).
11. Dirac, G. A.: Map Colour Theorems Related to the Heawood Colour Formula. J. London Math. Soc. **31**, 460–471 (1956).
12. Dirac, G. A.: On the Four-Color Conjecture. Proc. London Math. Soc. (3), **13**, 193–218 (1963).
13. Duke, R.: The Genus, Regional Number, and Betti Number of a Graph. Canad. J. Math. **18**, 817–822 (1966).
14. Dynkin, E. B., Uspenski, W. A.: Mehrfarbenprobleme (Mathematische Unterhaltungen I). Berlin: VEB Deutscher Verlag der Wissenschaften 1955 (translated from the Russian).
15. Edmonds, J.: A Combinatorial Representation for Polyhedral Surfaces. Notices Amer. Math. Soc. **7**, 646 (1960).
16. Erréra, A.: Du coloriage des cartes et de quelques questions d'Analysis situs. Paris and Brussels 1921.
17. Erréra, A.: Une vue d'ensemble sur le problème des quatre couleurs. Univ. Politec. Torino. Rend. Sem. Mat. **11**, 5–19 (1952).
18. Fraleigh, J. B.: A First Course in Abstract Algebra. Reading, Mass.: Addison-Wesley 1969. See pp. 365–368.
19. Franklin, P.: The Four Color Problem. Amer. J. Math. **44**, 225–236 (1922).
20. Franklin, P.: A Six Colour Problem. J. Math. Phys. **13**, 363–369 (1934).
21. Franklin, P.: Note on the Four Color Problem. J. Math. and Phys. **16**, 172 (1938).
22. Fréchet, M., Fan, Ky.: Initiation to Combinatorial Topology. Boston: Prindle, Weber and Schmidt 1967.

23. Garman, B.: private communication, 1973.
24. Graves, R.P.: Life of Sir William Rowan Hamilton. Volume 3. Dublin 1889.
25. Gross, J.L.: The Genus of Nearly Complete Graphs-Case 6. IBM Watson Research Center Preprint. New York: Yorktown Heights, 10pp.
26. Gross, J.L., Alpert, S.: Branched Coverings of Graph Imbeddings. IBM Research Pamphlet RC 4216. New York: Yorktown Heights 1973, 18pp.
27. Gross, J.L., Alpert, S.R.: The Topological Theory of Current Graphs. IBM Research Pamphlet, RC 4252. New York: Yorktown Heights 1973, 18pp.
28. Gustin, W.: Orientable Embedding of Cayley Graphs. Bull. Amer. Math. Soc. **69**, 272–275 (1963).
29. Guthrie, F.: Note on the Coloring of Maps. Proc. Roy. Soc. Edinburgh **10**, 727–728 (1880).
30. Guy, R.K.: A Combinatorial Problem, Nabla [=Bull. Malayan Math. Soc.], **7**, 68–72 (1960). (See also Dept. of Math., Univ. of Calgary, Res. Paper No. 8, 1967.)
31. Guy, R.K., Youngs, J.W.T.: A Smooth and Unified Proof of Cases 6, 5 and 3 of the Ringel-Youngs Theorem. J. Combinatorial Theory **15**, 1–11 (1973).
32. Guy, R.K., Jenkyns, T., Schaer, J.: The Toroidal Crossing Number of the Complete Graph. J. Combinatorial Theory **4**, 376–390 (1968).
33. Guy, R.K., Ringel, G.: Triangular Embedding of $K_n - K_6$. To appear.
34. Hadwiger, H.: Uber eine Klassifikation der Streckenkomplexe, Vierteljschr. Naturforsch. Ges. Zürich **88**, 133–142 (1943).
35. Heawood, P.J.: Map Colour Theorem. Quart. J. Math. **24**, 332–338 (1890).
36. Heesch, H.: Untersuchungen zum Vierfarbenproblem. B.I. Hochschulskripten, 810/810a/810b. Mannheim-Vienna-Zürich: Bibliographische Institute 1969.
37. Heffter, L.: Über das Problem der Nachbargebiete. Math. Ann. **38**, 477–508 (1891).
38. Hilbert, D., Cohn-Vossen, S.: Anschauliche Geometrie. Berlin: Springer 1932.
39. Himelwright, P.: On the Genus of Hamiltonian Groups, Specialist Thesis. Western Michigan University 1972.
40. Jacques, A.: Constellations et Propriétés Algèbraiques des Graphes Topologiques. Ph. D. Thesis. University of Paris 1969.
41. Jungerman, M.: Ph. D. Thesis. University of California, Santa Cruz, California 1974.
42. Kagno, I.N.: A note on the Heawood Color Formula. J. Math. Phys. **14**, 228–231 (1935).
43. Kainen, P.C.: A Lower Bound for Crossing Numbers of Graphs with Applications to K_n, $K_{p,q}$, and $Q(d)$. J. Combinatorial Theory **12**, 287–298 (1972).
44. Kainen, P.C.: Embeddings and Orientations of Graphs. In: Combinatorial Structures and their Applications, 193–196. New York: Gordon and Breach 1970.
45. Kempe, A.B.: On the Geographical Problem of the Four Colours. Amer. J. Math. **2**, 193–200 (1879).
46. Kempe, A.B.: How to Colour a Map with Four Colours. Nature **21**, 399–400 (1880).
47. Kleitman, D.J.: The Crossing Number of $K_{5,n}$. J. Combinatorial Theory **8**, 315–323 (1970).
48. König, D.: Theorie der endlichen und unendlichen Graphen. Kombinatorische Topologie der Streckenkomplexe (Mathematik und ihre Anwendungen, Bd. 16). Leipzig: Akad. Verlagsgesellschaft 1936.
49. Landesman, E.M., Youngs, J.W.T.: Smooth Solutions in Case 1 of the Heawood Conjecture for Non-orientable Surfaces. J. Combinatorial Theory **13**, 26–39 (1972).
50. Lang, S.: Algebra. Reading, Mass.: Addison-Wesley 1971. See pp. 182–185.
51. Mahnke, H.: Der kritische Fall 0 in der Heawoodschen Vermutung. Doctoral Thesis. Free University, Berlin 1971.
52. Mahnke, H.: The Necessity of Non-Abelian Groups in the Case 0 of the Heawood Map-Coloring Theorem. J. Combinatorial Theory **13**, 263–265 (1972).
53. May, K.O.: The Origin of the Four-Colour Conjecture. Isis **56**, 346–348 (1965).

54. Mayer, J.: Le Problème des Régions Voisines sur les Surfaces Closes Orientables. J. Combinatorial Theory **6**, 177–195 (1969).
55. Mayer, J.: Décomposition de K_{16} en Trois Graphes Planaires. J. Combinatorial Theory **13**, 71 (1972).
56. Nordhaus, E., Stewart, B., White, A.: On the Maximum Genus of a Graph. J. Combinatorial Theory **11**, 258–267 (1971).
57. Nordhaus, E. A., Stewart, B. M., Ringeisen, R. D., White, A.: A Kuratowski-type Theorem for the Maximum Genus of a Graph. J. Combinatorial Theory **B 12**, 260–267 (1972).
58. Ore, O.: Theory of Graphs. Amer. Math. Soc. Colloquium Publ. **XXXVIII**. Providence, R. I.: American Mathematical Society 1962.
59. Ore, O.: The Four Color Problem. Pure and Applied Mathematics **27**. New York: Acad. Pr. 1967.
60. Ore, O., Stemple, J.: On the Four Color Problem. Notices Amer. Math. Soc. **15**, No. 1, 196 (January 1968).
61. Petroelje, W. S.: Imbedding Graphs in Pseudosurfaces. Specialist Thesis. Western Michigan University 1971.
62. Reynolds, C. N.: On the Problem of Coloring Maps in Four Colours. Ann. of Math. **28**, 1 (1926).
63. Ringeisen, R. D.: Determining All Compact Orientable 2-Manifolds upon Which $K_{m,n}$ Has 2-Cell Imbeddings. J. Combinatorial Theory **12**, 101–104 (1972).
64. Ringeisen, R. D.: The Maximum Genus of a Graph. Ph. D. Thesis. Michigan State University 1970.
65. Ringel, G.: Farbensatz für nichtorientierbare Flächen beliebigen Geschlechts. J. reine angew. Math. **190**, 129–147 (1952).
66. Ringel, G.: Farbensatz für orientierbare Flächen vom Geschlecht $p>0$. J. reine angew. Math. **193**, 11–38 (1954).
67. Ringel, G.: Bestimmung der Maximalzahl der Nachbargebiete auf nichtorientierbaren Flächen. Math. Ann. **127**, 181–214 (1954).
68. Ringel, G.: Wie man die geschlossenen nichtorientierbaren Flächen in möglichst wenig Dreiecke zerlegen kann. Math. Ann. **130**, 317–326 (1955).
69. Ringel, G.: Über drei kombinatorische Probleme am n-dimensionalen Würfel und Würfelgitter. Abh. Math. Sem. Univ. Hamburg **20**, 10–19 (1955).
70. Ringel, G.: Das Geschlecht des vollständigen paaren Graphen. Abh. Math. Sem. Univ. Hamburg **28**, 139–150 (1965).
71. Ringel, G.: Färbungsprobleme auf Flächen und Graphen. Berlin: VEB Deutscher Verlag der Wissenschaften 1959.
72. Ringel, G.: Über das Problem der Nachbargebiete auf orientierbaren Flächen. Abh. Math. Sem. Univ. Hamburg **25**, 105–127 (1961).
73. Ringel, G., Youngs, J. W. T.: Solution of the Heawood Map-Coloring Problem. Proc. Nat. Acad. Sci. U.S.A. **60**, 438–445 (1968).
74. Ringel, G., Youngs, J. W. T.: Solution of the Heawood Map-Coloring Problem-Case 11. J. Combinatorial Theory **7**, 71–93 (1969).
75. Ringel, G., Youngs, J. W. T.: Solution of the Heawood Map-Coloring Problem-Case 8. J. Combinatorial Theory **7**, 353–363 (1969).
76. Ringel, G., Youngs, J. W. T.: Solution of the Heawood Map-Coloring Problem-Case 2. J. Combinatorial Theory **7**, 342–352 (1969).
77. Ringel, G., Youngs, J. W. T.: Lösung des Problems der Nachbargebiete auf orientierbaren Flächen. Archiv der Mathematik **20** (Basel), 190–201 (1969).
78. Ringel, G., Youngs, J. W. T.: Das Geschlecht des symmetrischen vollständigen dreifärbbaren Graphen. Comment. Math. Helv. **45**, 152–158 (1970).
79. Saaty, T. L.: The Minimum Number of Intersections in Complete Graphs. Proc. Nat. Acad. Sci. U.S.A. **52**, 688–690 (1964).

80. Sainte-Laguë, A.: Géométrie de situation et jeux. Mémorial Sci. Math., fasc. no 41. Paris: Gauthier-Villars 1929.
81. Seifert, H., Threlfall, W.: Lehrbuch der Topologie. Leipzig: Teubner 1934.
82. Tait, P.G.: Note on a Theorem in Geometry of Position. Trans. Roy. Soc. Edinburgh **29**, 657–660 (1880).
83. Tait, P.G.: Remarks on the Colouring of Maps. Proc. Roy. Soc. Edinburgh **10**, p. 729 (1880).
84. Terry, C.M., Welch, L.R., Youngs, J.W.T.: The Genus of K_{12s}. J. Combinatorial Theory **2**, 43–60 (1967).
85. Terry, C.M., Welch, L.R., Youngs, J.W.T.: Solution of the Heawood Map-Coloring Problem-Case 4. J. Combinatorial Theory **8**, 170–174 (1970).
86. Tietze, H.: Einige Bemerkungen über das Problem des Kartenfärbens auf einseitigen Flächen. Jahresber. Deutsch. Math. Vereinigung **19**, 155–159 (1910).
87. Turner, J.: Point Symmetric Graphs with a Prime Number of Points. J. Combinatorial Theory **3**, 136–145 (1967).
88. Waerden, B.L., van der: Modern Algebra, Volume 1. New York: Ungar 1953 [Translated by Fred Blum]. See pp. 115–118.
89. White, A.T.: The Genus of the Complete Tripartite Graph $K_{mn,n,n}$. J. Combinatorial Theory **7**, 283–285 (1969).
90. White, A.T.: The Genus of Repeated Cartesian Products of Bipartite Graphs. Trans. Amer. Math. Soc. **151**, 393–404 (1970).
91. White, A.T.: On the Genus of Products of Graphs, Recent Trends in Graph Theory (m. Capobianco, J.B. Frechen, and M. Krolik, editors). 217–219, Berlin: Springer 1971.
92. White, A.T.: The Genus of the Cartesian Product of Two Graphs. J. Combinatorial Theory **11**, 89–94 (1971).
93. White, A.T.: On the Genus of the Composition of Two Graphs. Pac. J. Math. **41**, 275–279 (1972).
94. White, A.T.: On the Genus of a Group. Trans. Amer. Math. Soc. **173**, 203–214 (1972).
95. Whitney, H., Tutte, W.T.: Kempe Chains and the Four Colour Problem. Utilitas Mathematica **2**, 241–281 (1972).
96. Winn, C.E.: A Case of Coloration in the Four Color Problem. Amer. J. Math. **29**, 515–528 (1937).
97. Winn, C.E.: On the Minimum Number of Polygons in an Irreducible Map. Amer. J. Math. **62**, 406–416 (1940).
98. Youngs, J.W.T.: Minimal Imbeddings and the Genus of a Graph. J. Math. Mech. **12**, 303–316 (1963).
99. Youngs, J.W.T.: The Heawood Map-Coloring Conjecture. Chapter 12 in Graph Theory and Theoretical Physics (F. Harary, ed.), 313–354. New York, London: Academic Press 1967.
100. Youngs, J.W.T.: Remarks on the Heawood Conjecture (Nonorientable Case). Bull. Amer. Math. Soc. **74**, 347–353 (1968).
101. Youngs, J.W.T.: The Nonorientable Genus of K_n. Bull. Amer. Math. Soc. **74**, 354–358 (1968).
102. Youngs, J.W.T.: Solution of the Heawood Map-Coloring Problem-Cases 3, 5, 6, and 9. J. Combinatorial Theory **8**, 175–219 (1970).
103. Youngs, J.W.T.: The Heawood Map-Coloring Problem-Cases 1, 7, and 10. J. Combinatorial Theory **8**, 220–231 (1970).
104. Zaks, J.: The Maximum Genus of Cartesian Products of Graphs. To appear.

Index

Die Grundlehren der mathematischen Wissenschaften in Einzeldarstellungen mit besonderer Berücksichtigung der Anwendungsgebiete

Eine Auswahl

---

23. Pasch: Vorlesungen über neuere Geometrie
41. Steinitz: Vorlesungen über die Theorie der Polyeder
45. Alexandroff: Topologie. Band 1
46. Nevanlinna: Eindeutige analytische Funktionen
63. Eichler: Quadratische Formen und orthogonale Gruppen
102. Nevanlinna/Nevanlinna: Absolute Analysis
114. Mac Lane: Homology
127. Hermes: Enumerability, Decidability, Computability
131. Hirzebruch: Topological Methods in Algebraic Geometry
135. Handbook for Automatic Computation. Vol. 1/Part a: Rutishauser: Description of ALGOL 60
136. Greub: Multilinear Algebra
137. Handbook for Automatic Computation. Vol. 1/Part b: Grau/Hill/Langmaack: Translation of ALGOL 60
138. Hahn: Stability of Motion
139. Mathematische Hilfsmittel des Ingenieurs. 1. Teil
140. Mathematische Hilfsmittel des Ingenieurs. 2. Teil
141. Mathematische Hilfsmittel des Ingenieurs. 3. Teil
142. Mathematische Hilfsmittel des Ingenieurs. 4. Teil
143. Schur/Grunsky: Vorlesungen über Invariantentheorie
144. Weil: Basic Number Theory
145. Butzer/Berens: Semi-Groups of Operators and Approximation
146. Treves: Locally Convex Spaces and Linear Partial Differential Equations
147. Lamotke: Semisimpliziale algebraische Topologie
148. Chandrasekharan: Introduction to Analytic Number Theory
149. Sario/Oikawa: Capacity Functions
150. Iosifescu/Theodorescu: Random Processes and Learning
151. Mandl: Analytical Treatment of One-dimensional Markov Processes
152. Hewitt/Ross: Abstract Harmonic Analysis. Vol. 2: Structure and Analysis for Compact Groups. Analysis on Locally Compact Abelian Groups
153. Federer: Geometric Measure Theory
154. Singer: Bases in Banach Spaces I
155. Müller: Foundations of the Mathematical Theory of Electromagnetic Waves
156. van der Waerden: Mathematical Statistics
157. Prohorov/Rozanov: Probability Theory
159. Köthe: Topological Vector Spaces
160. Agrest/Maksimov: Theory of Incomplete Cylindrical Functions and their Applications
161. Bhatia/Shegö: Stability Theory of Dynamical Systems
162. Nevanlinna: Analytic Functions
163. Stoer/Witzgall: Convexity and Optimization in Finite Dimensions
164. Sario/Nakai: Classification Theory of Riemann Surfaces
165. Mitrinović/Vasić: Analytic Inequalities
166. Grothendieck/Dieudonné: Eléments de Géométrie Algébrique I

167. Chandrasekharan: Arithmetical Functions
168. Palamodov: Linear Differential Operators with Constant Coefficients
170. Lions: Optimal Control Systems Governed by Partial Differential Equations
171. Singer: Best Approximation in Normed Linear Spaces by Elements of Linear Subspaces
172. Bühlmann: Mathematical Methods in Risk Theory
173. F. Maeda/S. Maeda: Theory of Symmetric Lattices
174. Stiefel/Scheifele: Linear and Regular Celestial Mechanics. Perturbed Two-body Motion—Numerical Methods—Canonical Theory
175. Larsen: An Introduction of the Theory of Multipliers
176. Grauert/Remmert: Analytische Stellenalgebren
177. Flügge: Practical Quantum Mechanics I
178. Flügge: Practical Quantum Mechanics II
179. Giraud: Cohomologie non abélienne
180. Landkoff: Foundations of Modern Potential Theory
181. Lions/Magenes: Non-Homogeneous Boundary Value Problems and Applications I
182. Lions/Magenes: Non-Homogeneous Boundary Value Problems and Applications II
183. Lions/Magenes: Non-Homogeneous Boundary Value Problems and Applications III
184. Rosenblatt: Markov Processes. Structure and Asymptotic Behavior
185. Rubinowicz: Sommerfeldsche Polynommethode
186. Wilkinson/Reinsch: Handbook for Automatic Computation II. Linear Algebra
187. Siegel/Moser: Lectures on Celestial Mechanics
188. Warner: Harmonic Analysis on Semi-Simple Lie Groups I
189. Warner: Harmonic Analysis on Semi-Simple Lie Groups II
190. Faith: Algebra: Rings, Modules, and Categories I
192. Mal'cev: Algebraic Systems
193. Pólya/Szegö: Problems and Theorems in Analysis. Vol. 1
194. Igusa: Theta Functions
195. Berberian: Baer *-Rings
196. Athreya/Ney: Branching Processes
197. Benz: Vorlesungen über Geometrie der Algebren
198. Gaal: Linear Analysis and Representation Theory
200. Dold: Lectures on Algebraic Topology
203. Schoeneberg: Elliptic Modular Functions
206. André: Homologie des algèbres commutatives
207. Donoghue: Monotone Matrix Functions and Analytic Continuation
209. Ringel: Map Color Theorem